Oxford Mathematics K

Primary Years Programme

Contents

NUMBER, PATTERN AND FUNCTION

Unit 1 Number and place value
1. Numbers 0 to 10 — 2
2. Counting to 10 — 6
3. How many? — 10
4. Numbers without counting
5. Comparing numbers — 18
6. Ordinal numbers — 22
7. Numbers 10 to 20 — 26
8. Teen numbers — 30
9. More than and less than — 34
10. Adding to 10 — 38
11. Grouping — 42
12. Sharing — 46

Unit 2 Fractions and decimals
1. Halves — 50

Unit 3 Money and financial mathematics
1. Money — 54

Unit 4 Patterns and algebra
1. Sorting — 58
2. Repeating patterns — 62
3. Creating and describing patterns — 66

MEASUREMENT, SHAPE AND SPACE

Unit 5 Using units of measurement
1. Length, height and area — 70
2. Volume and capacity — 75
3. Mass — 80
4. — 84
5. Days of the week — 88

Unit 6 Shape
1. 2D shapes — 92
2. 3D shapes — 96

Unit 7 Location and transformation
1. Position — 100
2. Directions — 104

DATA HANDLING

Unit 8 Data representation and interpretation
1. Yes or no questions — 108
2. Pictographs — 112

Glossary — 116
Answers — 124

OXFORD
UNIVERSITY PRESS
AUSTRALIA & NEW ZEALAND

UNIT 1: TOPIC 1
Numbers 0 to 10

Count on.

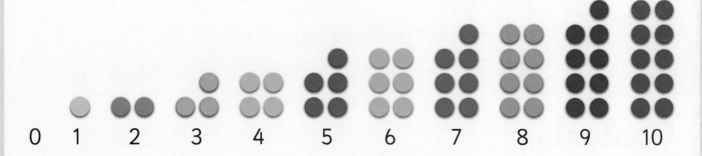

Count back.

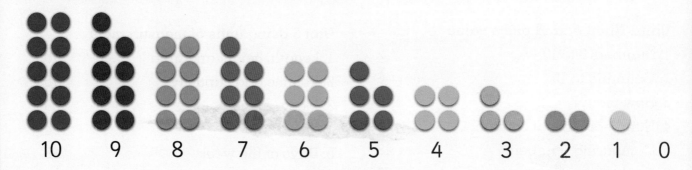

What number comes after 10?

Guided practice

1. Trace the numbers.

a

0 1 2 3 4 5 6 7 8 9 10

b

10 9 8 7 6 5 4 3 2 1 0

Independent practice

1 Copy the numbers.

a

0	1	2	3	4	5	6	7	8	9	10

b

10	9	8	7	6	5	4	3	2	1	0

2 Fill in the gaps.

a

0	1	2			5				9	

b

10	9	8		6			3			

c

4	5	6				10	

3 What comes after?

a 1 ☐   b 2 ☐   c 4 ☐

d 5 ☐   e 7 ☐   f 9 ☐

Which number is the biggest?

4 What comes before?

a ☐ 2   b ☐ 4   c ☐ 5

d ☐ 7   e ☐ 9   f ☐ 10

5 What comes before and after?

a ☐ 2 ☐   b ☐ 8 ☐

c ☐ 5 ☐   d ☐ 7 ☐

e ☐ 6 ☐   f ☐ 9 ☐

Extended practice

1

a Number the carriages 0 to 4.

b Draw the matching number of people.

---

2 Circle the group of:

a 3.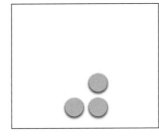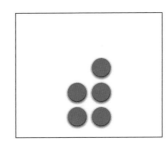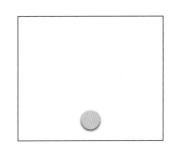

b 6.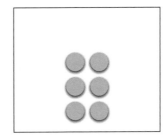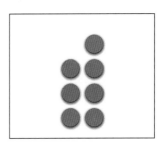

c 8.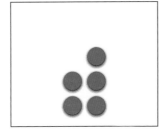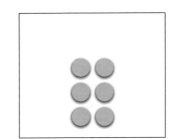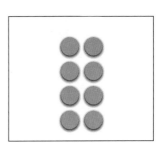

UNIT 1: TOPIC 2
Counting to 10

1 cat

4 cats

7 cats

Which box has the most cats in it?

Guided practice

1 Match the number to the objects.

a

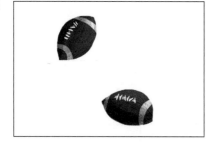      

| 4 | 2 | 9 |

b

| 7 | 5 | 6 |

Independent practice

1 Count how many.

a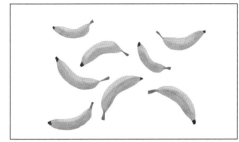

b

c

d

e

f

g

h

2 Draw:

a   7 pencils.

b   3 stars.

c   5 squares.

d   10 circles.

e   4 books.

f   6 faces.

Extended practice

1 How many dots?

a 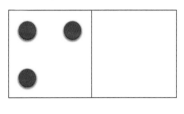   b 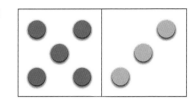   c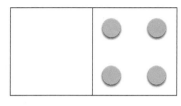

2 Draw dots to match the numbers.

a 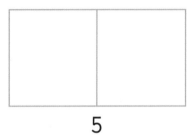   b    c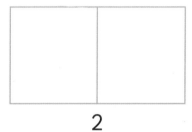

5   9   2

3 Order from smallest to biggest.

a   4   7   3   5

b   10   1   6   8

What's the biggest number you know?

UNIT 1: TOPIC 3
How many?

   is the same number as

4   4

If I take away 1 hat, which group will have more?

Guided practice

1 a How many?

 ☐   🐕🐕🐕🐕🐕 ☐

b Are the groups the same size?   Yes   No

---

2 a How many?

 ☐    ☐

b Are the groups the same size?   Yes   No

Independent practice

1 Circle the matching number.

a  4 5 6

b  7 8 9

c  6 8 10

d  6 8 10

e 3 4 5

> Remember to count to check you have the right number.

2 Draw more to make:

a 3.

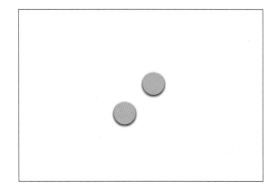

b 5.

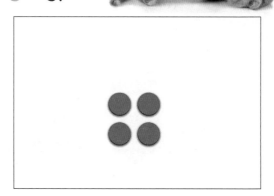

c 8.

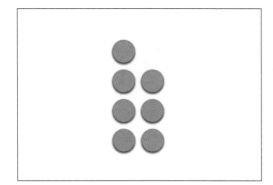

d 4.

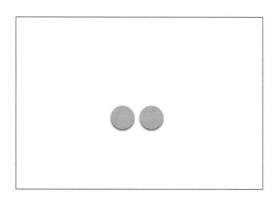

e 6.

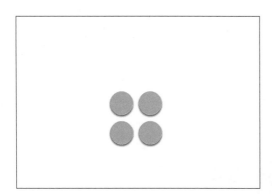

f 10.

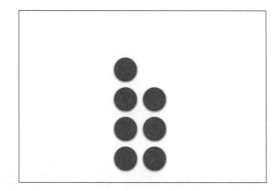

g 9.

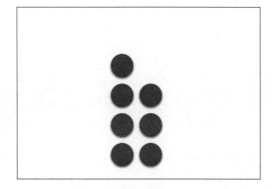

h 8.

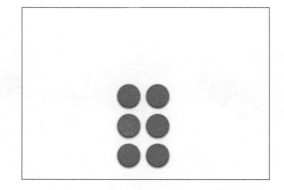

Extended practice

1 Draw lines to match groups of the same number.

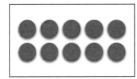

2

a How many?

b Draw 1 more. How many now?

3

a How many?

b Draw 2 more. How many now?

UNIT 1: TOPIC 4
Numbers without counting

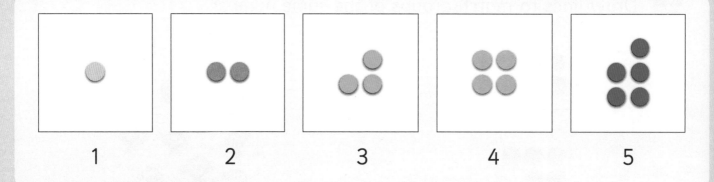

Guided practice

1 Write how many without counting.

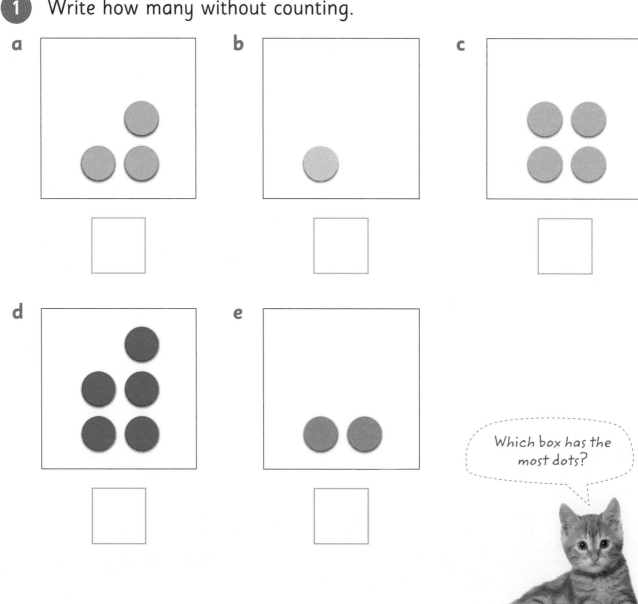

Which box has the most dots?

Independent practice

1 Match the numbers and pictures without counting.

a

| 3 | 1 | 2 | 3 | 2 |

b

| 3 | 2 | 5 | 4 | 6 |

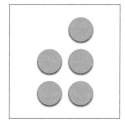 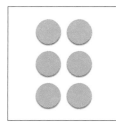  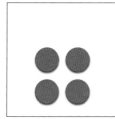 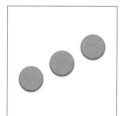

c

| 7 | 3 | 4 | 5 | 10 |

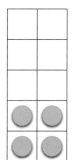 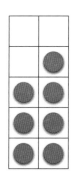   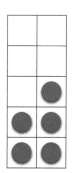

2 Write how many dots without counting.

a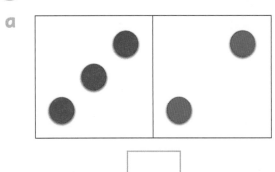

b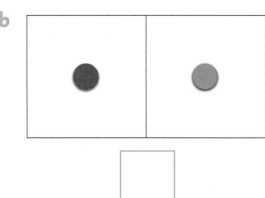

c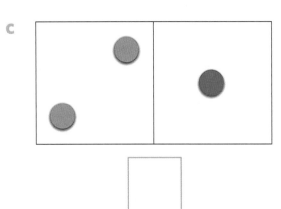

d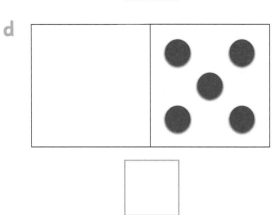

e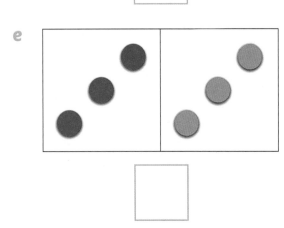

f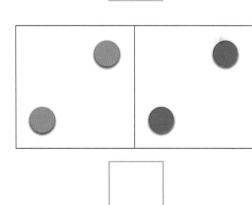

3 Circle the odd group out.

a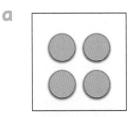

Hmm. Do all the groups have the same number of items in them?

b

Extended practice

1 Redraw from smallest to largest.

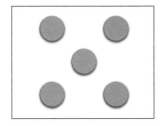 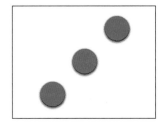 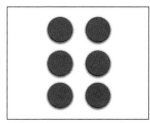 

---

2 Draw lines to match groups of the same number.

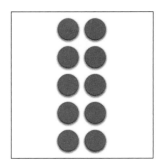 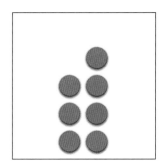  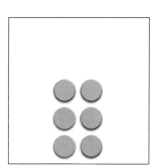

UNIT 1: TOPIC 5
Comparing numbers

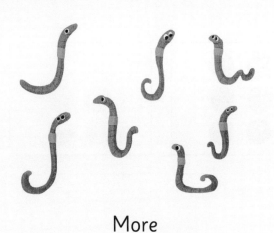

More

Less

How do you know which group has more?

Guided practice

1 Circle the group with more.

a

b

c

Independent practice

1. a How many? ☐   b Draw a group with more.

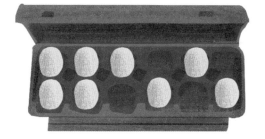 

2. a How many? ☐   b Draw a group with more.

3. a How many? ☐   b Draw a group with less.

4. a How many? ☐   b Draw a group with less.

5 Does the second group have less, more or the same number?

Can you work out the answers without counting?

a

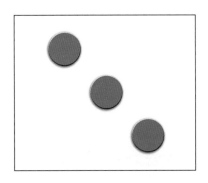 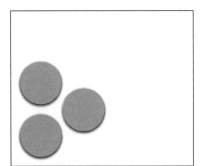

Less
More
Same

b

Less
More
Same

c

Less
More
Same

d

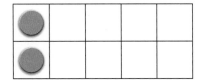 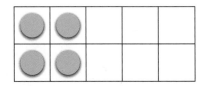

Less
More
Same

Extended practice

1 Match the pictures, numbers and ten frames.

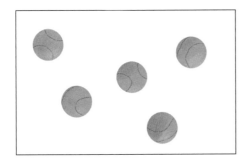 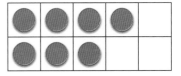  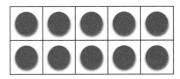  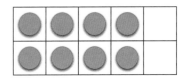 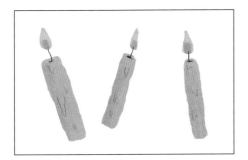   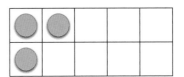

UNIT 1: TOPIC 6
Ordinal numbers

1st	2nd	3rd	4th	5th	6th
first	second	third	fourth	fifth	sixth

When have you been first at something?

Guided practice

1 Follow the instructions to colour the mice.

a 1st: **red**   b 2nd: **grey**   c 3rd: **purple**
d 4th: **blue**   e 5th: **yellow**   f 6th: **green**

2 What colour is:

a the 1st?

b the 2nd?

c the 6th?

Independent practice

1 Match the words and numbers.

| first | second | third | fourth | fifth | sixth |

| 3rd | 1st | 6th | 5th | 2nd | 4th |

2 Label the birds from 1st to 6th.

3 Rewrite in the correct order.

| second | fourth | third | first |

| | | | |

4 Look at the picture.

Which animal is:

a 1st?

b 6th?

c second?

d third?

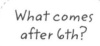

What comes after 6th?

---

5 Circle the:

a 2nd.

b 5th.

c 4th.

Extended practice

1 Match the activities to their order.

| 4th | 1st | 3rd | 2nd |

---

2 Draw a:

a in the 1st box.   b in the third box.

c in the 6th box.   d in the last box.

e  in the 4th box.   f  in the second box.

UNIT 1: TOPIC 7
Numbers 10 to 20

Count on.

10 11 12 13 14 15 16 17 18 19 20

Count back.

20 19 18 17 16 15 14 13 12 11 10

What number comes after 20?

Guided practice

1 Trace the numbers.

a Count on.

10 11 12 13 14 15 16 17 18 19 20

b Count on.

5 6 7 8 9 10 11 12 13 14 15

c Count back.

20 19 18 17 16 15 14 13 12 11 10

Independent practice

1 Copy the numbers.

a

10	11	12	13	14	15	16	17	18	19	20

b

20	19	18	17	16	15	14	13	12	11	10

2 Fill in the gaps.

a

| 10 | 11 | | 13 | 14 | | | 17 | 18 | | |

b

| 20 | | 18 | 17 | 16 | | | 13 | 12 | | 10 |

c

| 13 | 14 | | 16 | 17 | | |

3. **What comes after?**

    a 10 ☐   b 13 ☐   c 15 ☐

    d 16 ☐   e 18 ☐   f 19 ☐

How many digits are in each of the numbers?

4. **What comes before?**

    a ☐ 11   b ☐ 13   c ☐ 10

    d ☐ 16   e ☐ 19   f ☐ 20

5. **What comes before and after?**

    a ☐ 14 ☐   b ☐ 17 ☐

    c ☐ 15 ☐   d ☐ 12 ☐

    e ☐ 19 ☐   f ☐ 11 ☐

Extended practice

1 Circle the group of:

a 13.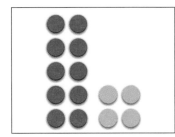

b 18.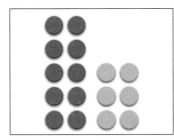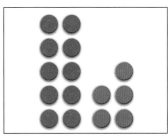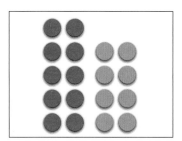

2 Fill in the missing numbers.

11	12	13			16		18		20
21		23	24	25		27		29	30

3 Write the next 2 numbers.

a 9, __, __   b 18, __, __

c 21, __, __   d 27, __, __

e 16, __, __   f 26, __, __

UNIT 1: TOPIC 8
Teen numbers

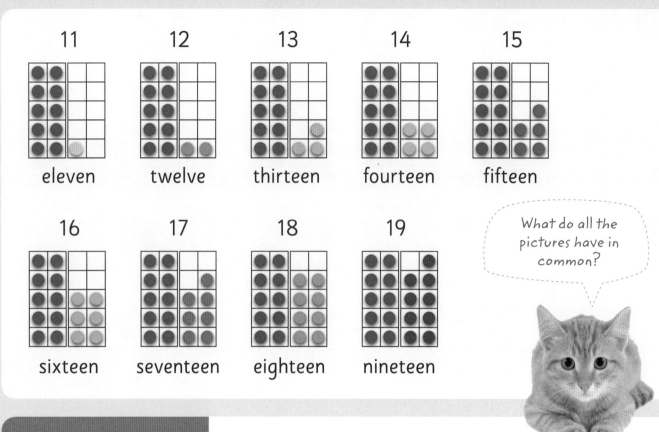

Guided practice

1 Match the numbers to the groups.

a

11   15   16

b

18   13   17

Independent practice

1 Match the words and numbers.

a

| eleven | twelve | thirteen | fourteen | fifteen |

| 14 | 13 | 11 | 15 | 12 |

b

| sixteen | seventeen | eighteen | nineteen |

| 16 | 19 | 17 | 18 |

2 Count how many.

a

b

c

d

3
a How many **blue**?
b How many **red**?
c How many altogether?

4
a How many **blue**?
b How many **yellow**?
c How many altogether?

5
a How many **blue**?
b How many **purple**?
c How many altogether?

6
a How many **blue**?
b How many **teal**?
c How many altogether?

What would 20 look like on the ten frames?

Extended practice

1 Draw more to make:

a 12.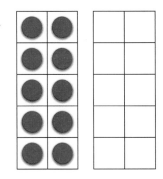

b 17.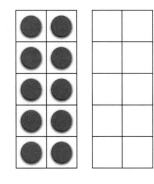

c 18.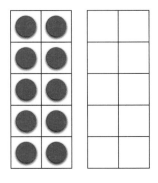

d 14.

2 Order from smallest to largest.

a | 19 | 12 | 2 | 14 | 16 | 5 |

| | | | | 16 | |

b | 20 | 17 | 10 | 6 | 11 | 7 |

| | | | | | |

c | 15 | 5 | 12 | 2 | 11 | 1 |

| | | | | | |

UNIT 1: TOPIC 9
More than and less than

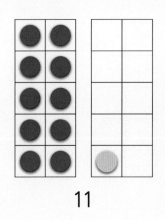
11

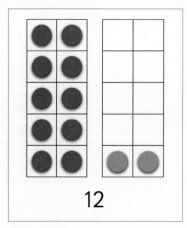
12

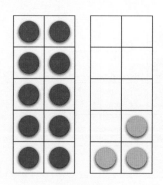
13

12 is 1 **more** than 11.

12 is 1 **less** than 13.

Is 11 more or less than 13?

Guided practice

1 Circle:

a 1 **more** than 13.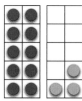

b 1 **more** than 15.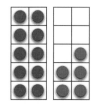

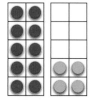 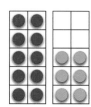 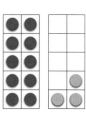

c 1 **less** than 13.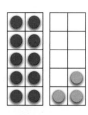

d 1 **less** than 16.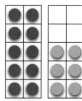

 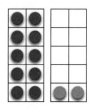 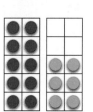

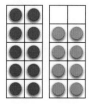 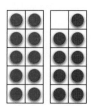 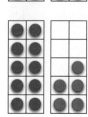

Independent practice

1 Draw a group with **more**.

a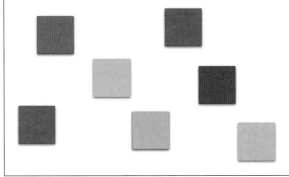

b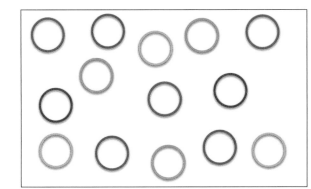

2 Draw a group with **less**.

a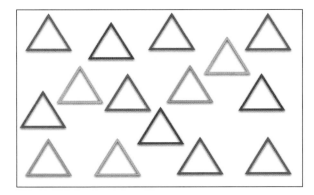

b

3 a Circle the group with **more**.

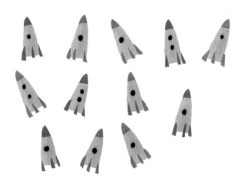 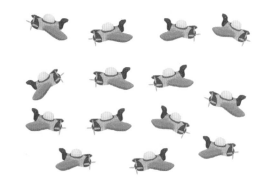

b How many more?

4 a Circle the group with **more**.

b How many more?

5 a Circle the group with **less**.

b How many less?

Which group on this page has the most items in it?

Extended practice

1

a Are there more than 12 animals? | Yes | No |

b Are there less than 20 animals? | Yes | No |

c Are there more dogs than cats? | Yes | No |

d Are there more than 12 cats? | Yes | No |

2 a Draw 3 more.   b How many now? ☐

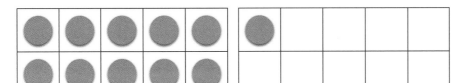

3 a Draw 5 more.   b How many now? ☐

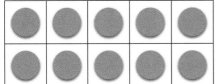 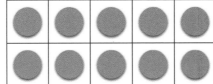 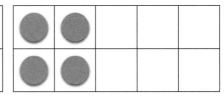

UNIT 1: TOPIC 10
Adding to 10

Pairs that make 4

0 and 4 makes 4

1 and 3 makes 4

2 and 2 makes 4

3 and 1 makes 4

4 and 0 makes 4

When you join 2 numbers together, it is called adding.

Guided practice

1 Record the pairs that make 5.

a

☐ and ☐ makes 5

b

☐ and ☐ makes 5

c

☐ and ☐ makes 5

d

☐ and ☐ makes 5

e

☐ and ☐ makes 5

f

☐ and ☐ makes 5

Independent practice

1 Draw lines to match the pairs that make 7.

 3    1    5    2    4

 2    3    4    6    5

2 Draw lines to match the pairs that make 10.

 5    8    6    3    1

 7    2    9    5    4

Can you work out how many items are in each box without counting?

3 Fill in the gaps.

a) ☐ and ☐ makes ☐

b) ☐ and ☐ makes ☐

c) ☐ and ☐ makes ☐

d) ☐ and ☐ makes ☐

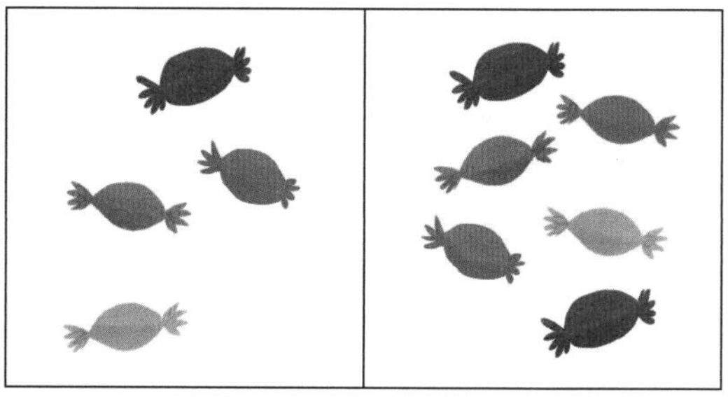

e) ☐ and ☐ makes ☐

f) ☐ and ☐ makes ☐

g) ☐ and ☐ makes ☐

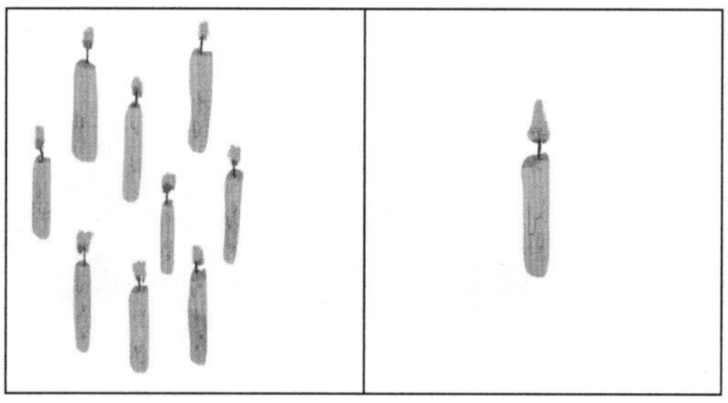

h) ☐ and ☐ makes ☐

Extended practice

1 How many more to make 10?

a
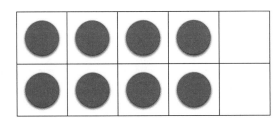

b
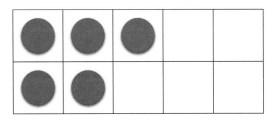

c
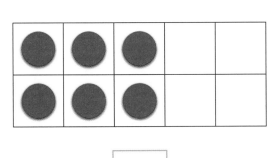

d
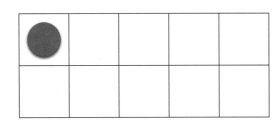

---

2 a Draw pencils in the jars so there are 10 altogether.
b Fill in the gaps to match your picture.

  ☐ and ☐ makes ☐

UNIT 1: TOPIC 11
Grouping

6 koalas

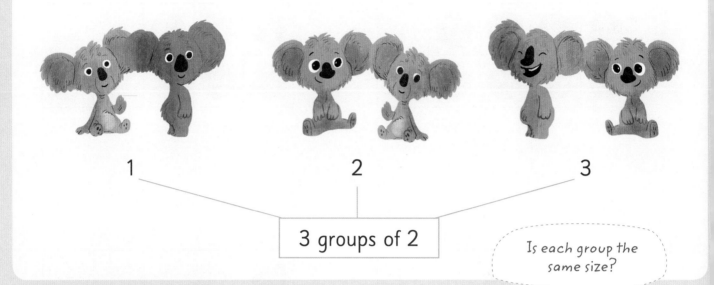

3 groups of 2

Is each group the same size?

Guided practice

1. a How many altogether? ☐

   b How many groups of 2? ☐

2. a How many altogether? ☐

   b How many groups of 3? ☐

Independent practice

1 Draw circles to make groups of:

a 2.

b 3.

c 1.

d 2.

e 5.

2 Are the groups equal or unequal?

a 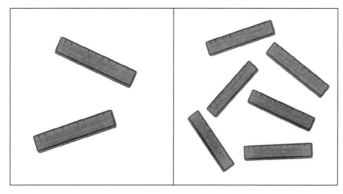 Equal / Unequal

b 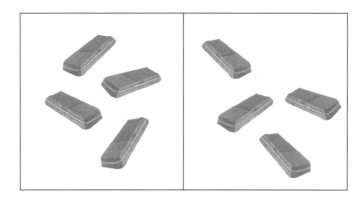 Equal / Unequal

c 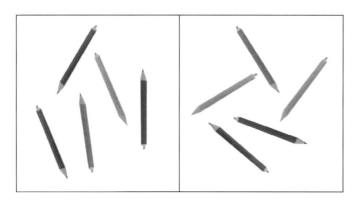 Equal / Unequal

d 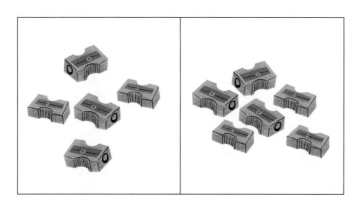 Equal / Unequal

How many more would you need to make the groups equal?

Extended practice

1 a Draw circles to make equal groups.

b How many in each group? ☐

---

2 a Draw circles to make equal groups.

b How many in each group? ☐

---

3 a Draw circles to make equal groups.

b How many in each group? ☐

UNIT 1: TOPIC 12
Sharing

Sharing

2 people

4 cookies

2 cookies each

The shares need to be equal.

Guided practice

1 Draw how many each person gets.

a

4 people

4 cookies

b

3 people

6 cookies

Independent practice

1   **a**   Share the fish into the ponds.

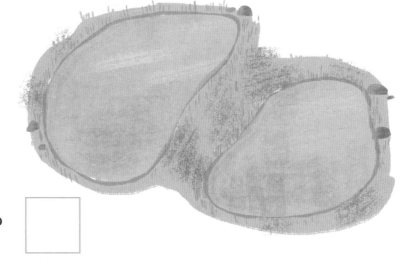

    b   How many in each? ☐

2   **a**   Share the jellybeans into the jars.

    b   How many in each? ☐

3   **a**   Share the balls into the boxes.

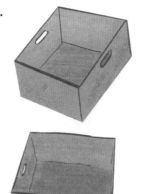

    b   How many in each? ☐

4 a Draw lines to share 6 bananas between 2 monkeys.

b How many each?

5 a Draw lines to share 10 muffins between 5 children.

Could 10 muffins be shared fairly between 4 children?

b How many each?

6 a Draw lines to share 9 sheep between 3 pens.

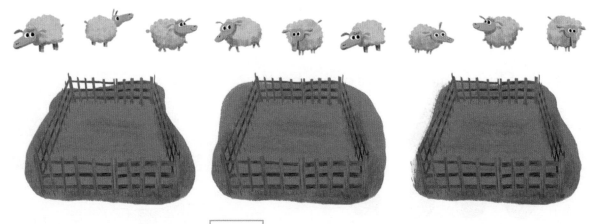

b How many in each?

Extended practice

1 **a** Draw more balloons to make the shares fair.

  b How many each? ☐

  c How many altogether? ☐

2 **a** Draw more doughnuts to make the shares fair.

  b How many each? ☐

  c How many altogether? ☐

3 Tick the fair shares.

a

b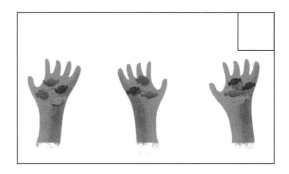

c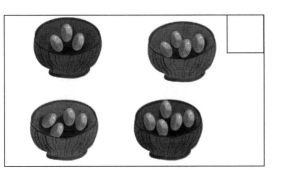

d

UNIT 2: TOPIC 1
Halves

1 whole

1 half

1 half

How many halves make a whole?

Guided practice

1 Colour half.

a

b
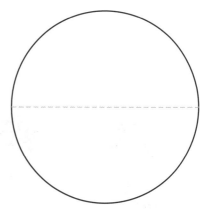

c
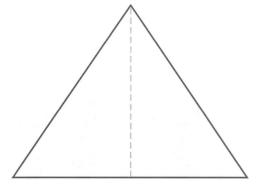

d
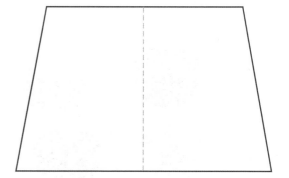

Independent practice

1 Half or whole?

a

| Half | Whole |

b

| Half | Whole |

c

| Half | Whole |

d

| Half | Whole |

e

| Half | Whole |

f

| Half | Whole |

2 Circle the picture that shows halves.

a

Halves need to be the same size.

b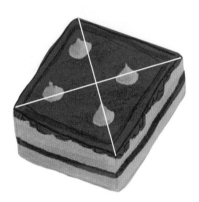

c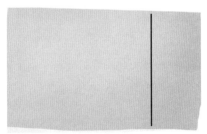

d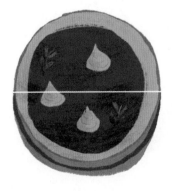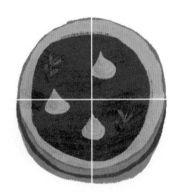

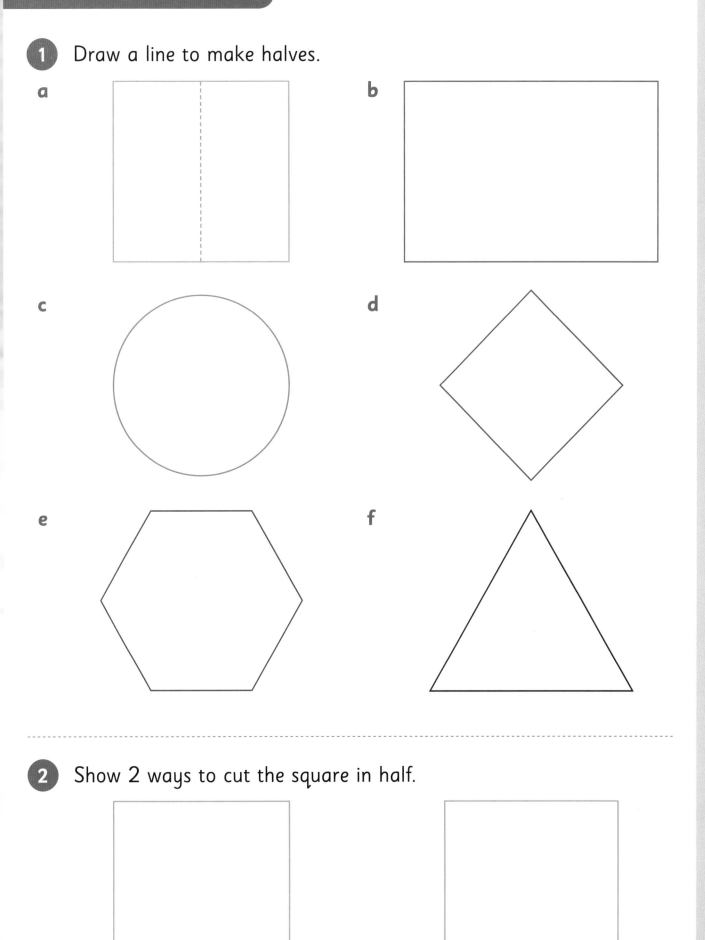

UNIT 3: TOPIC 1
Money

We use money to buy all sorts of things.

food   clothes   books   pets

Coins and notes are also called cash.

Guided practice

1 Tick the places where you would need money.

at the movies   walking in the park   at the supermarket

☐   ☐   ☐

asleep in bed   at an ice-cream shop   in a restaurant

☐   ☐   ☐

Independent practice

1 Ava dropped all her coins. How many did she have?

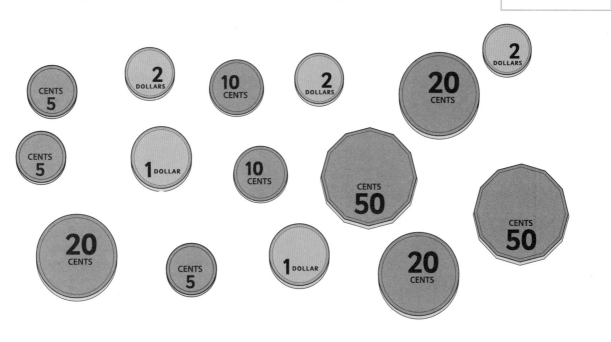

2 Ben emptied his money box.

a How many coins does he have?

b How many notes does he have?

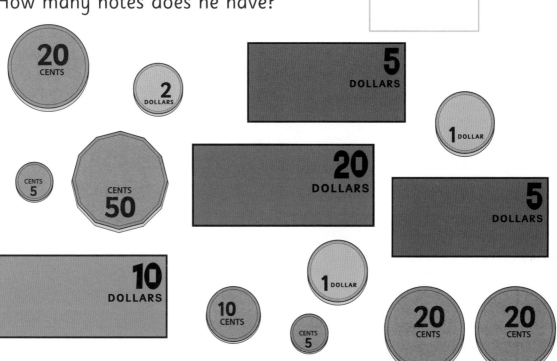

3 Flynn and Hannah compared their coin collections.

Flynn   Hannah

a How many does Flynn have?

b How many does Hannah have?

c Who has more? | Flynn | Hannah |

Flynn and Hannah's coins are from all around the world!

4 Draw lines to match the coins that are the same.

Extended practice

1 Draw 2 different places you might need money.

2 Ellie went shopping. Circle the coins she would need to buy the following:

a

b

c

UNIT 4: TOPIC 1
Sorting

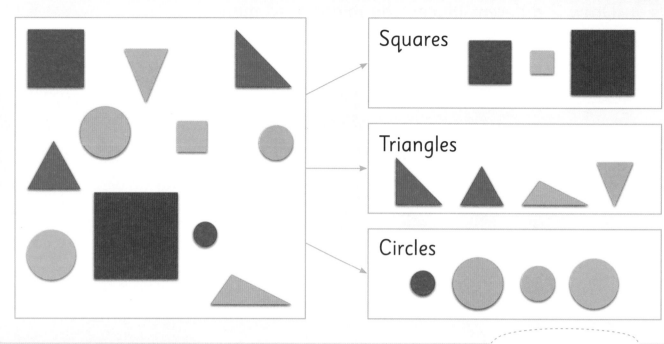

Guided practice

1 a Colour the squares **red**.

b Colour the triangles **blue**.

c Colour the circles **green**.

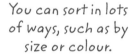

You can sort in lots of ways, such as by size or colour.

2 How many:

a  ?

b ?

c ?

Independent practice

1 a Circle all the small shapes.

  b Colour the rectangles **blue**.

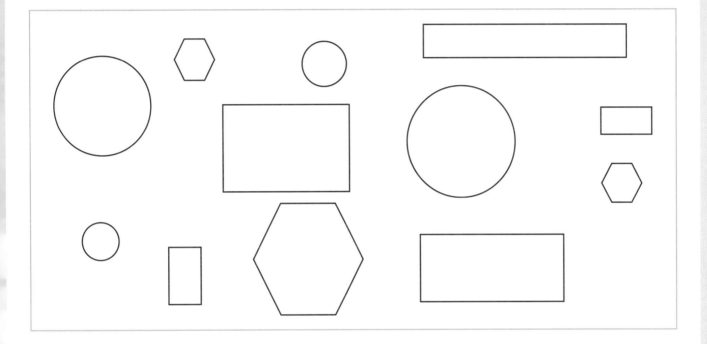

2 a Colour all the clothes **red**.

  b Colour all the animals **brown**.

  c Circle all the bottles.

3) Circle the item that doesn't belong.

a

b

c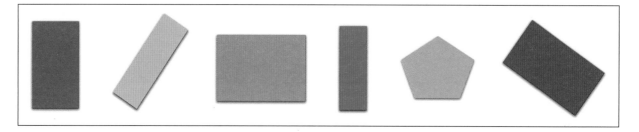

4) Draw lines to sort the items.

| People | Dogs | Birds |

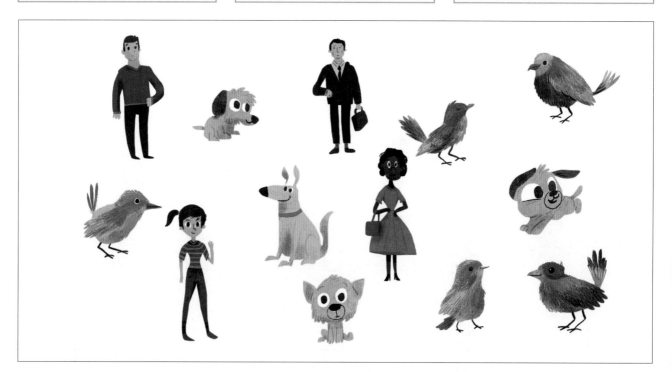

Extended practice

1 Sort the items into groups by colouring.

Could you sort the items another way?

---

2 Draw the shapes sorted into groups.

UNIT 4: TOPIC 2
Repeating patterns

Shape pattern

Picture pattern

The rule for the picture pattern is cat, cat, dog.

Guided practice

1 Continue the patterns.

a

b

c

d

Independent practice

1 Complete the patterns.

a

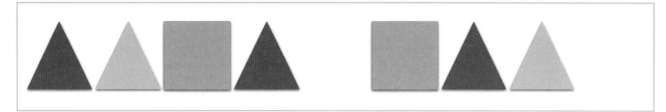

b

c

d

e

2 Draw the part of the pattern that repeats.

a

b

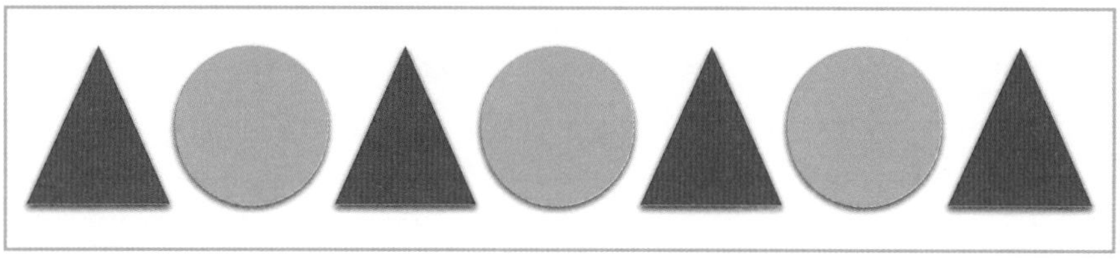

c

3 Finish the colour patterns.

a

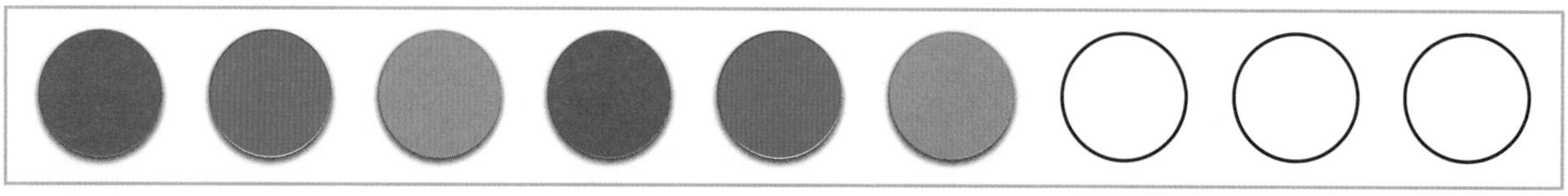

b

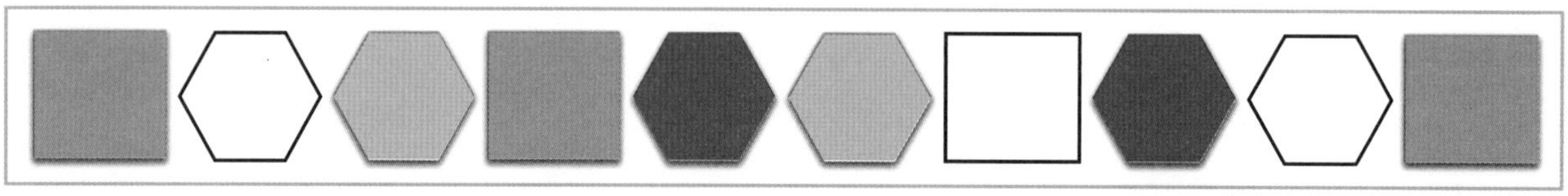

c

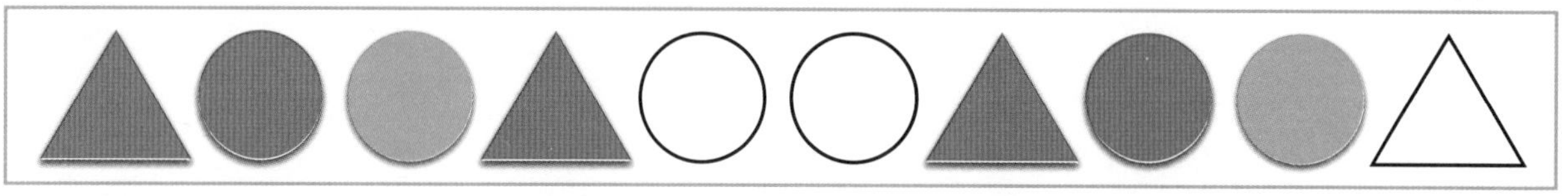

Where can you find patterns?

Extended practice

1 Use the shapes to make a pattern.

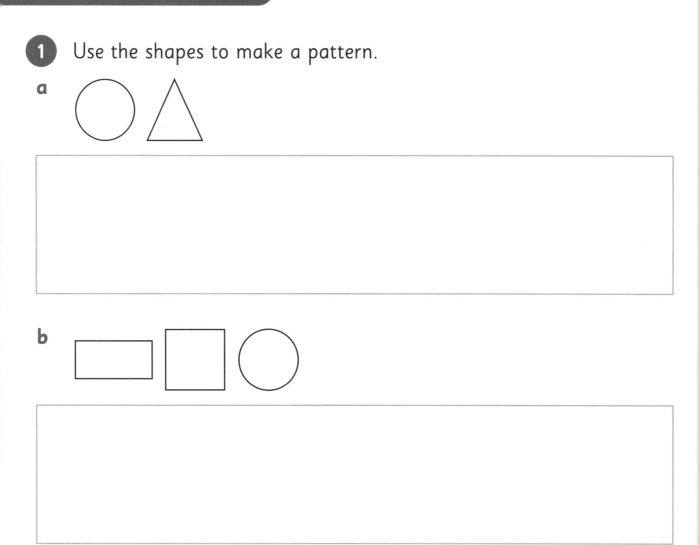

2 Circle the shape that doesn't belong in the pattern.

a

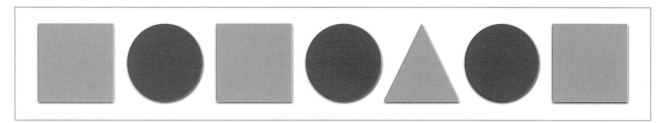

b

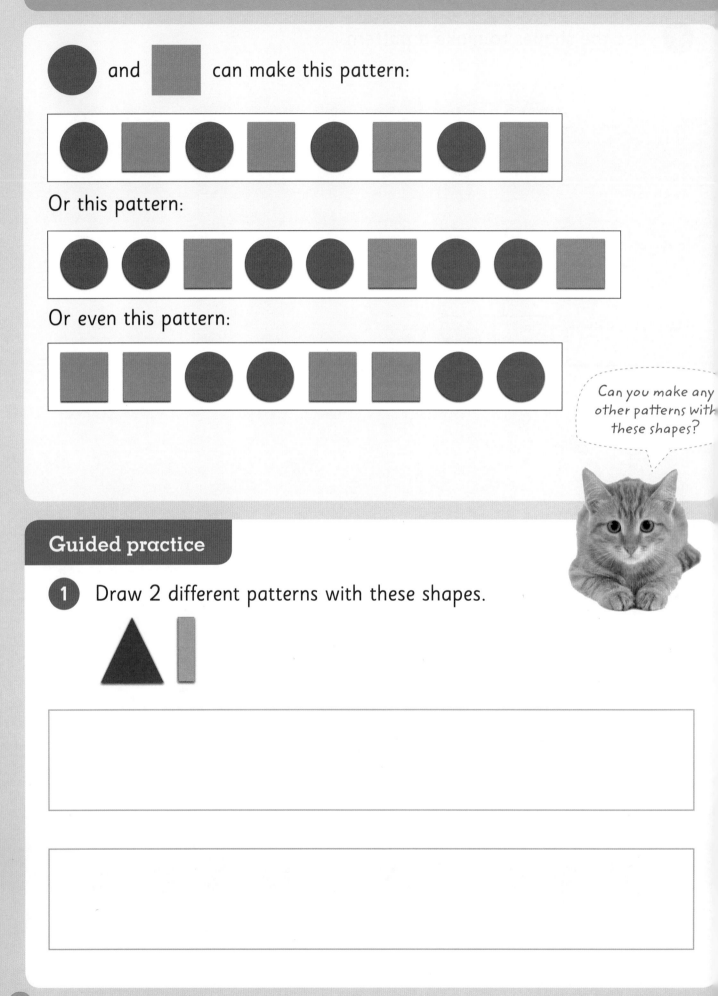

Independent practice

1 a Circle the part of the pattern that repeats.

b How many items did you circle?

2 a Circle the part of the pattern that repeats.

b How many items did you circle?

3 a Circle the part of the pattern that repeats.

b How many items did you circle?

4 a Circle the part of the pattern that repeats.

b How many items did you circle?

5 a Circle the mistake.

b Draw the correct shape.

6 a Circle the mistake.

b Draw the correct shape.

How did you know which shape to draw?

7 Circle what comes next.

a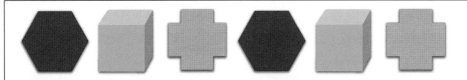

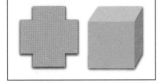 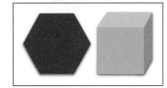 

b

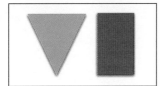 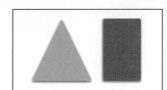 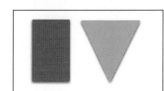

Extended practice

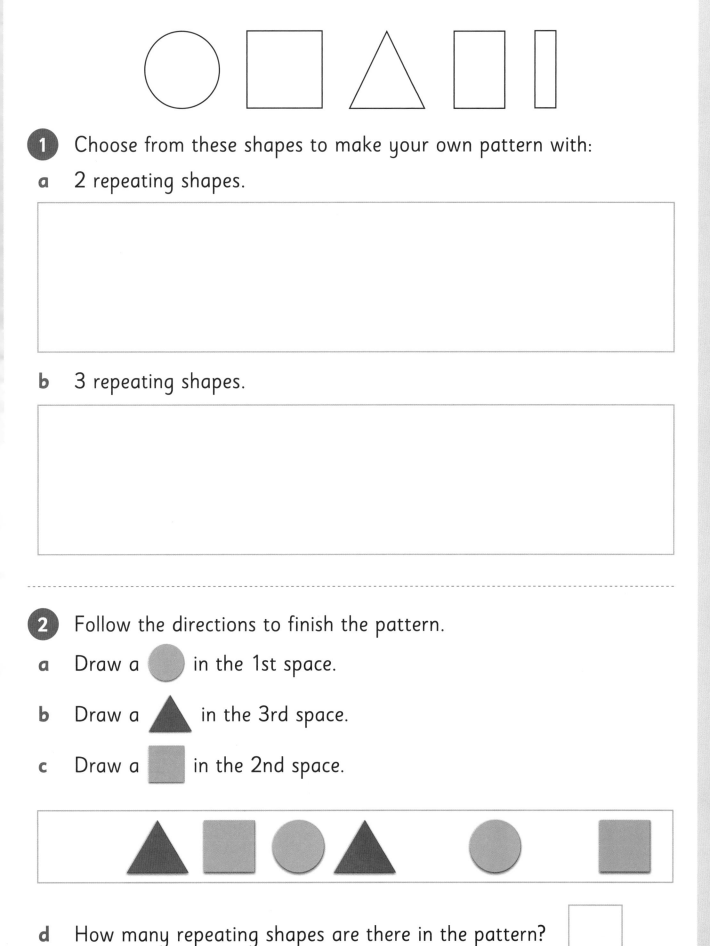

1 Choose from these shapes to make your own pattern with:

a 2 repeating shapes.

b 3 repeating shapes.

2 Follow the directions to finish the pattern.

a Draw a ⬤ in the 1st space.

b Draw a ▲ in the 3rd space.

c Draw a ■ in the 2nd space.

d How many repeating shapes are there in the pattern?

UNIT 5: TOPIC 1
Length, height and area

Length

Longer

Shorter

Taller　　Shorter

You need to line the ends of objects up to check which object is longer.

Guided practice

1 Circle the **longer** item.

a

b

2 Circle the **shorter** item.

a

b

Independent practice

1 Draw a line that is:

a longer.

b shorter.

2 Draw a building that is:

a taller.

b shorter.

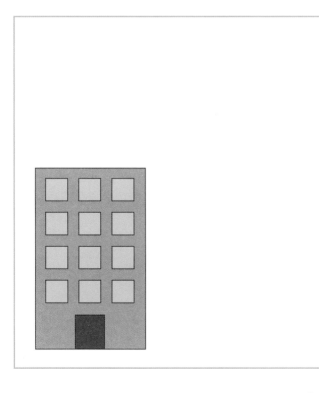
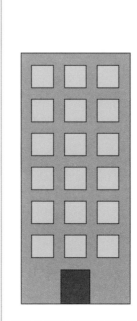

Area

The surface of the blue book is bigger than the green book.

The blue book has a bigger area.

You can place one object on top of another to compare their areas.

Guided practice

1 Circle the shape with the bigger area.

a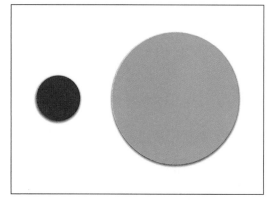

b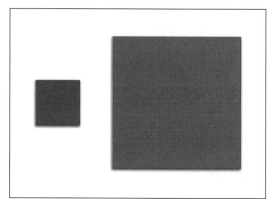

c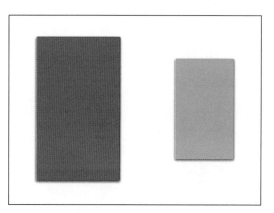

d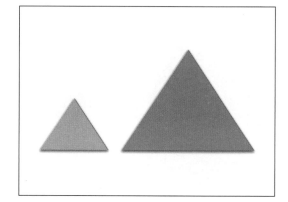

Independent practice

1 a Cover the front of a book with square blocks.

How many did you need? ☐

b Cover the front of a different book with square blocks.

How many did you need? ☐

c Draw the book with the bigger area.

Try putting one of your books on top of the other to see which has a bigger area.

2 Find and draw an object with an area of about 8 blocks.

3 Find and draw an object with an area of about 12 blocks.

Extended practice

1 Find and draw something:

a longer than your hand span.

b shorter than your hand span.

c taller than you.

2 Find and draw something with an area:

a bigger than this book.

b smaller than this book.

UNIT 5: TOPIC 2
Volume and capacity

Volume

Takes up more space   Takes up less space

The amount of space an item takes up is called its volume.

Guided practice

1 Circle the item that takes up **more** space.

a

b

2 Circle the item that takes up **less** space.

a

b

Independent practice

1 Draw something that takes up **more** space than:

a

b

2

a Make this object with blocks.

b Use the same blocks to make this object.

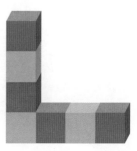

c Circle the object that needed **more** blocks.

Capacity

Holds more

Holds less

The amount a container can hold is called its capacity.

Guided practice

1 Circle the container that holds **more**.

a

b

2 Circle the container that holds **less**.

a

b

Independent practice

1 Draw something that holds **more** than:

a

b

---

2 Match the words and pictures.

| Full | Half-full | Empty |

Extended practice

1
 a Choose 2 empty containers.
 b Fill with water to find which container holds more.
 c Draw the containers.

| Holds more | Holds less |

2
 a Choose 2 empty boxes.
 b Fill with blocks to find out which box takes up more space.
 c Draw the boxes.

| Takes up more space | Takes up less space |

UNIT 5: TOPIC 3
Mass

Heavy

Light

You can lift objects with your hands to see if they are heavy or light. This is called hefting.

Guided practice

1 Circle the **heavier** object.

a

b

2 Circle the **lighter** object.

a

b

Independent practice

1 Heft each pair of objects. Circle the **heavier** one.

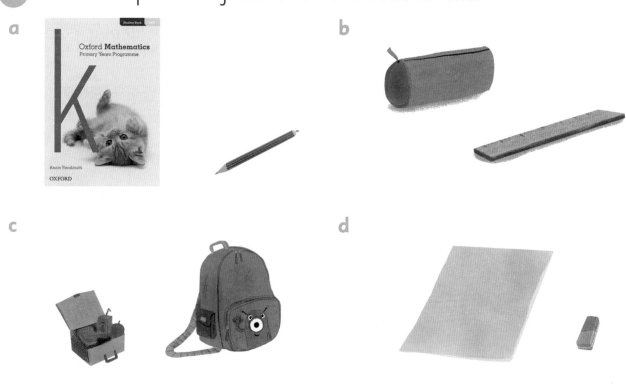

2 Heft each pair. Circle the **lighter** one.

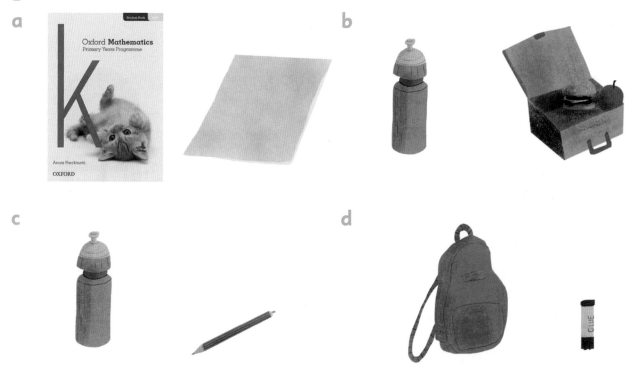

3 Heavy or light?

a

| Heavy | Light |

b

| Heavy | Light |

c

| Heavy | Light |

Are bigger objects always heavier?

4 Colour the easiest object to push or pull.

a

b

5 Colour the hardest object to push or pull.

a

b

Extended practice

1 Draw something **heavier** than:

a

b

2 Draw something **lighter** than:

a

b

3 a Draw an object that is easy to push or pull.

b Draw an object that is hard to push or pull.

UNIT 5: TOPIC 4
Time

I get up at 7 o'clock.

I start school at 9 o'clock.

I get home at 4 o'clock.

For o'clock times, the big hand is on the 12 and the little hand tells you what hour it is.

Guided practice

1 Circle the event that happens first.

a

b

2 Draw lines to match the clocks and the times.

| 7 o'clock | 5 o'clock | 10 o'clock |

Independent practice

1 Draw lines to show the order.

| 1st | 2nd | 3rd | 4th |

2

a Circle the things that take a **long** time.

b Circle the things that take a **short** time.

3) What time is it?

a

☐ o'clock

b

☐ o'clock

c

☐ o'clock

d

☐ o'clock

e

☐ o'clock

f

☐ o'clock

4) Tick the things that happen at night.

Extended practice

1 Draw something that you do:

a in the morning.

b in the afternoon.

---

2 About what time do you:

a go to bed?   **b** eat lunch?   **c** go to school?

 o'clock   o'clock   o'clock

UNIT 5: TOPIC 5
Days of the week

Monday   Tuesday | Wednesday | Thursday   Friday   **Saturday**   **Sunday**

If today is Wednesday, yesterday was Tuesday and tomorrow will be Thursday.

There are 5 weekdays and 2 weekend days.

Guided practice

1 a Write the days of the week in the right order.

Tuesday	Monday
Monday	
Wednesday	
Friday	
Saturday	
Sunday	
Thursday	

b Write your favourite day of the week. _____

Independent practice

1 a Colour the weekdays **blue**.

b Colour the weekend days **red**.

| Monday | Tuesday | Wednesday | Thursday | Friday | Saturday | Sunday |

2 Draw lines to show:

a the days you go to school.

b the days you have sport at school.

c what day it is today.

d the day after Saturday.

| Monday |

| Tuesday |

| Wednesday |

| Thursday |

| Friday |

| Saturday |

| Sunday |

3 Draw something you do on:

a Monday.

b Friday.

c Saturday.

d Sunday.

4 a Today is ⬚.

b Yesterday was ⬚.

c Tomorrow will be ⬚.

Days of the week always start with a capital letter.

Extended practice

1 Fill in the missing days.

a | Monday | | Wednesday | |

b | | Sunday | Monday | |

> What smaller word is at the end of all the days of the week?

2 What day comes **after**:

a Tuesday?

b Friday?

c Thursday?

d Sunday?

3 What day comes **before**:

a Monday?

b Saturday?

c Wednesday?

d Friday?

UNIT 6: TOPIC 1
2D shapes

 Circle

 Triangle

 Square

 Rectangle

2D stands for two-dimensional. 2D shapes are flat.

Guided practice

1 a Colour the circles **red**.

   b Colour the squares **green**.

2 a Colour the rectangles **blue**.

   b Colour the triangles **yellow**.

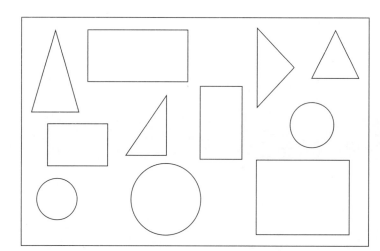

Independent practice

1 Draw lines to match the shapes that are the same.

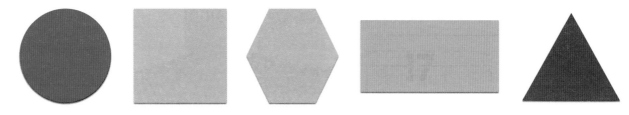

2 a Trace the straight lines in **blue**.
b Trace the curved lines in **red**.

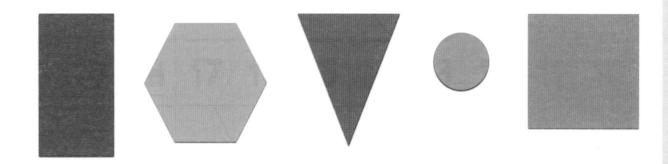

3 How many:

a   corners? ☐   b   corners? ☐
    sides? ☐   sides? ☐

c   corners? ☐   d   corners? ☐
    sides? ☐   sides? ☐

A side is a straight line that joins one corner of a shape to another corner.

4 How many:

a   triangles? ☐   b   rectangles? ☐

c   circles? ☐

Extended practice

1 Draw a shape with:

a 4 corners and 4 sides.

b no corners and no sides.

c 3 corners and 3 sides.

2 How many:

a squares?

b circles?

c triangles?

UNIT 6: TOPIC 2
3D shapes

Sphere

Cube

Cylinder

Cone

3D stands for three-dimensional. 3D shapes are not flat.

Guided practice

1 a Circle the spheres in **red**.

b Circle the cubes in **green**.

2 a Circle the cylinders in **blue**.

b Circle the cones in **yellow**.

Independent practice

1 Match the drawings with the real objects.

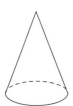

2 a Draw the 2D shape that makes up the faces of a cube.

b How many sides does a cube have?

③ Circle the objects that can roll.

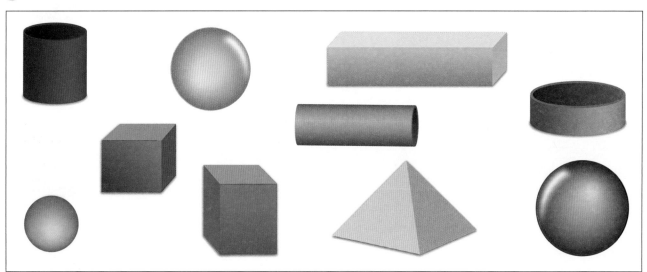

④ How many:

a spheres? ☐

b cubes? ☐

c cylinders? ☐

d cones? ☐

Objects that can roll have at least one curved face.

Extended practice

1 Circle the object that does not belong.

a

b

c

2 a Draw some objects that are cubes.

b Draw some objects that are spheres.

UNIT 7: TOPIC 1
Position

The tree is **next to** the house.

The boy is **under** the tree.

The dog is **near** the boy.

What other words can describe where something is?

Guided practice

1 Colour the word that describes the position of each item.

a The is [under / above] the .

b The is moving [towards / away from] the .

c The   are [above / below] the .

d The 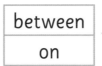 is [between / on] the and the .

Independent practice

1 Draw a ball:

a **inside** the box.

b **on** the table.

c **next to** the bat.

d **between** the boy and the girl.

e **near** the cat.

f **under** the car.

2 Left or right?

a  The dog is to the [left/right] of the cat.

b  The cheese is to the [left/right] of the bread.

c  The book is to the [left/right] of the hat.

Do you write with your left or right hand?

3

a Draw a tree **behind** the cow.

b Draw a boy **next to** the cow.

c Draw a balloon **above** the boy.

d Draw a hat **on** the boy.

Extended practice

1 Fill in the blanks.

a The clock is [............] the books.

b The teacher is [............] the table.

c The board is [............] the teacher.

---

2 True or false?

a The books are **beneath** the globe. ☐ True ☐ False

b The apple is **between** the teacher and the books. ☐ True ☐ False

c The clock is **inside** the bookcase. ☐ True ☐ False

UNIT 7: TOPIC 2
Directions

To get from to  :

Walk **along** the path.
Turn **right**.
Walk **down** to the bus stop.

How would you get from the house to the pond?

Guided practice

1 a Draw a **red** path from the to the .

  b Draw a **blue** path from the to the .

  c Draw a **green** path from the to the .

Independent practice

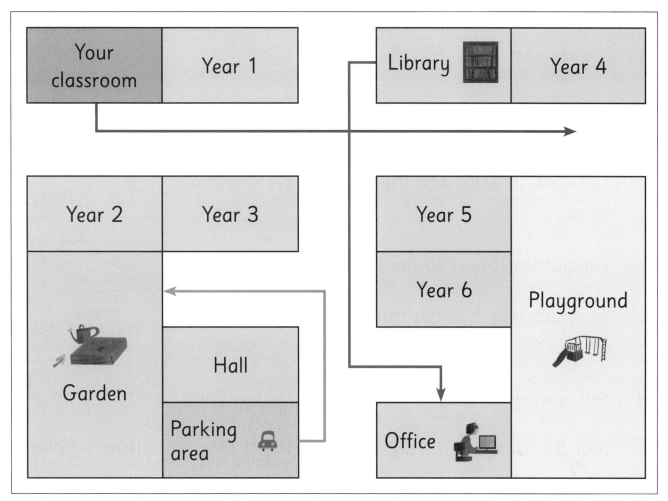

① Follow the **blue** path.

a It starts at [your classroom | the hall].

b It ends near [Year 2 | Year 4].

c It goes past [the parking area | Year 3].

② Follow the **green** path.

a It starts at [Year 6 | the parking area].

b It ends at [the library | the garden].

c It goes past [the hall | Year 1].

4 Follow the **red** path.

a It starts at [the library | your classroom].

b It ends at [the garden | the office].

c It goes past [Year 4 | Year 5].

What is your real classroom near?

4 True or false?

a You need to walk past the office to get from the library to Year 3. — True / False

b The garden is next to the hall. — True / False

c You pass Year 6 to get from the parking area to your classroom. — True / False

d You can't get from the garden to the playground. — True / False

e Year 3 is closer to the hall than Year 4. — True / False

5 Start at your classroom.

a Turn right between Year 3 and Year 5.

b Turn left between Year 6 and the office.

c Keep walking. Where do you end up?

Extended practice

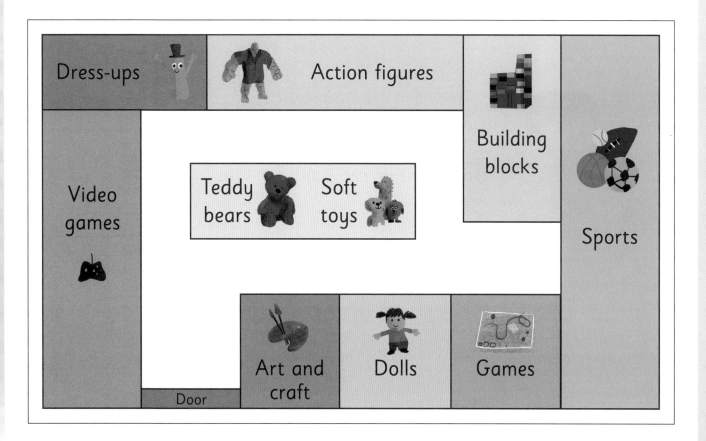

1 **a** Draw a path from the Door to your favourite place in the toy store.

  b Circle the places your path goes past.

  c Draw a different way to get to your favourite place.

  d Circle the places your new path goes past.

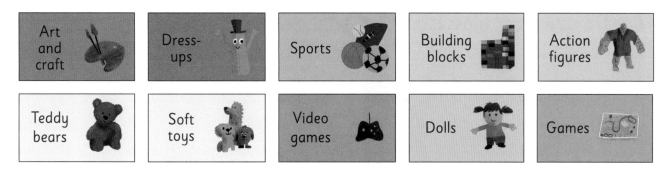

UNIT 8: TOPIC 1
Yes or no questions

	Yes	No
Are you a girl?	**Yes**	No
Do you own a dog?	**Yes**	No
Do you have red hair?	Yes	**No**

What other yes/no questions could you ask the girl?

Guided practice

1 Answer these questions.

a Are you a boy?  | Yes | No |

b Do you own a dog? | Yes | No |

c Do you have brown hair? | Yes | No |

d Do you wear glasses? | Yes | No |

e Are you at school?  | Yes | No |

Independent practice

1   a   Ask 5 people if they like bananas.
Record their answers in the table.

Person 1		Person 2		Person 3		Person 4		Person 5	
yes	no	yes	no	yes	no	yes	no	yes	no

  b   How many said yes? ☐

  c   How many said no? ☐

2   a   Ask 5 people if they have ever been on a plane. Record their answers in the table.

Person 1		Person 2		Person 3		Person 4		Person 5	
yes	no	yes	no	yes	no	yes	no	yes	no

  b   How many said yes? ☐

  c   How many said no? ☐

3   a   Ask 5 people if they walked to school today.
Record their answers in the table.

Person 1		Person 2		Person 3		Person 4		Person 5	
yes	no	yes	no	yes	no	yes	no	yes	no

  b   How many said yes? ☐

  c   How many said no? ☐

4 Look at the chart.

Question: Do you like football?

Yes	
No	

a How many said yes? ☐

b How many said no? ☐

c Did more say yes or no? ☐

Would you have answered yes or no?

5 Colour the faces to match the answers.

Question: Are you 5 years old?

Person 1	Person 2	Person 3	Person 4	Person 5
yes no	yes **no**	yes no	**yes** no	yes **no**

Yes	
No	

a How many said yes? ☐

b How many said no? ☐

Extended practice

1 Write 2 questions you could ask your class.

a Do you have _____?

b Do you like _____?

2

a Choose one of your questions. Write it here.

_____

b Ask 5 people your question.

Person 1		Person 2		Person 3		Person 4		Person 5	
yes	no	yes	no	yes	no	yes	no	yes	no

c Colour the faces to match their answers.

Yes	☺	☺	☺	☺	☺
No	😐	😐	😐	😐	😐

UNIT 8: TOPIC 2
Pictographs

My marbles

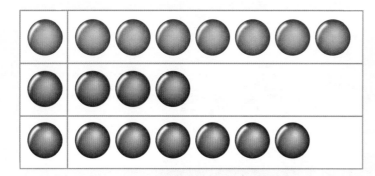

I have 7 **green** marbles.
I have 3 **red** marbles.
I have 6 **blue** marbles.

How many marbles does he have altogether?

Guided practice

Favourite juices in our class

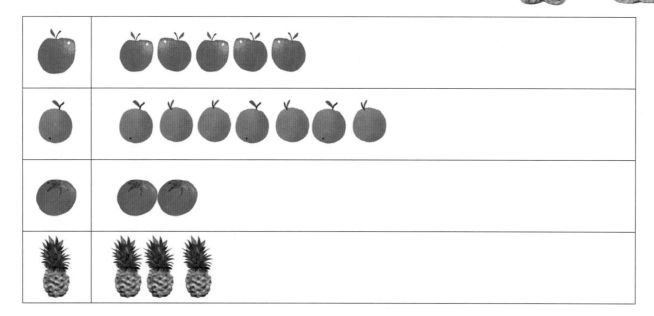

1 Look at the graph.

a How many ? □   b How many  ? □

c How many ? □   d How many   ? □

Independent practice

1 Look at the graph.

Favourite treats

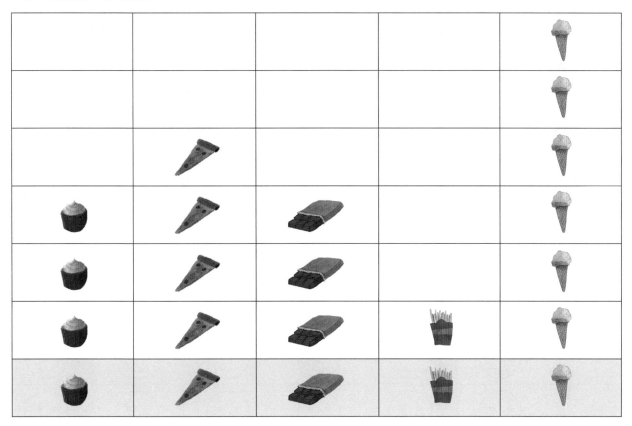

a Circle the **most popular** treat.

b Circle the **least popular** treat.

c Do more people like than ? | Yes | No |

What is your favourite treat?

2

Can you swim?

a   Colour the graph to match the answers.

Can you swim?

Yes	☺	☺	☺	☺	☺	☺
No	😐	😐	😐	😐	😐	😐

b   Did more people say yes? | Yes | No |

c   How many people said no?   ☐

d   How many people said yes? ☐

e   How many people were asked? ☐

Extended practice

1 a Ask 10 people what their favourite colour is.

b Colour the graph to show their responses.

Favourite colours

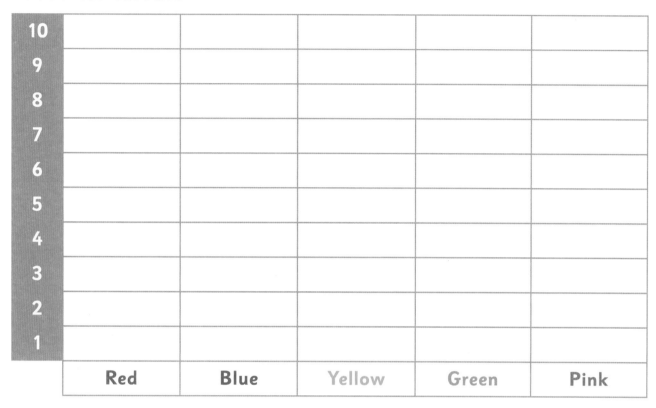

2 Use your graph to answer these questions.

a How many like **red**?

b How many like **blue**?

c How many like **pink**?

d Circle the most popular colour.

   red   blue   yellow   green   pink

e Circle the least popular colour.

   red   blue   yellow   green   pink

Is your favourite colour the most popular?

GLOSSARY

addition The joining or adding of two numbers together to find the total. Also known as *adding*, *plus* and *sum*.

Example:

★★★ + ★★ = ★★★★★

3 and 2 is 5

anticlockwise Moving in the opposite direction to the hands on a clock.

area The size of an object's surface.

Example:
It takes 12 tiles to cover this placemat.

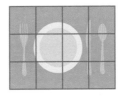

array An arrangement of items into even columns and rows that make them easier to count.

balance scale Equipment that balances items of equal mass – used to compare the mass of different items. Also called pan balance or equal arm balance.

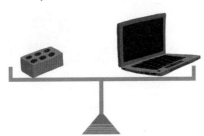

base The bottom edge of a 2D shape or the bottom face of a 3D shape.

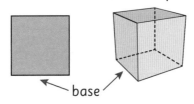

calendar A chart or table showing the days, dates, weeks and months in a year.

	January 2017					
Sun	Mon	Tues	Wed	Thur	Fri	Sat
1	2	3	4	5	6	7
8	9	10	11	12	13	14
15	16	17	18	19	20	21
22	23	24	25	26	27	28
29	30	31				

Month → January 2017 ← Year
Day ↗
Date →

capacity The amount that a container can hold.

Example:
The jug has a capacity of 4 cups.

4 cups
3 cups
2 cups
1 cup

category A group of people or things sharing the same characteristics.

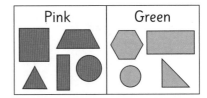

centimetre A unit for measuring the length of smaller items.

Example: Length is 15 cm.

80 cm

circle A 2D shape with a continuous curved line that is always the same distance from the centre point.

clockwise Moving in the same direction as the hands on a clock.

cone A 3D shape with a circular base that tapers to a point.

corner The point where two edges of a shape or object meet.

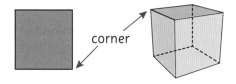

cube A rectangular prism where all 8 faces are squares of equal size.

cylinder A 3D shape with 2 parallel circular bases and one curved surface.

data Information gathered through methods such as questioning, surveys or observation.

day A period of time that lasts 24 hours.

difference (between) A form of subtraction or take away.

Example: The difference between 11 and 8 is 3.

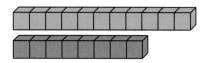

digit The single numerals from 0 to 9. They can be combined to make larger numbers.

Example: 24 is a 2-digit number. 378 is a 3-digit number.

division/dividing Sharing into equal groups.

Example: 9 divided by 3 is 3

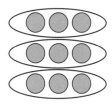

double/doubles Adding two identical numbers or multiplying a number by 2.

Example: 4 + 4 = 8   2 x 4 = 8

duration How long something lasts.

Example: The school week lasts for 5 days.

edge The side of a shape or the line where two faces of an object meet.

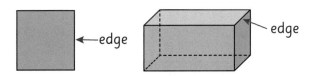

eighth One part of a whole or group divided into eight equal parts.

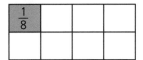

Eighth of a whole   Eighth of a group

equal Having the same number or value.

Example:

Equal size   Equal numbers

equation A written mathematical problem where both sides are equal.

Example: 4 + 5 = 6 + 3

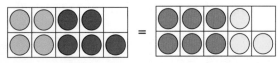

estimate A thinking guess.

face The flat surface of a 3D shape.

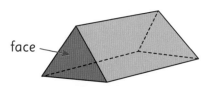

flip To turn a shape over horizontally or vertically. Also known as reflection.

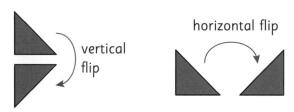

fraction An equal part of a whole or group.

Example: One out of two parts or $\frac{1}{2}$ is shaded.

friendly numbers Numbers that are easier to add to or subtract from.

Example: 10, 20 or 100

half One part of a whole or group divided into two equal parts. Also used in time for 30 minutes.

Example:

 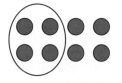

Half of   Half of   Half past 4
a whole   a group

hexagon A 2D shape with 6 sides.

horizontal Parallel with the horizon or going straight across.

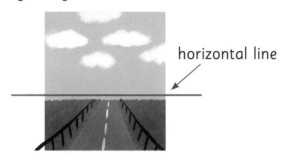

horizontal line

jump strategy A way to solve number problems that uses place value to "jump" along a number line by hundreds, tens and ones.

Example: 16 + 22 = 38

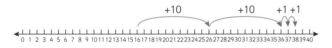

length How long an object is from end to end.

Example: This poster is 3 pens long.

mass How heavy an object is.

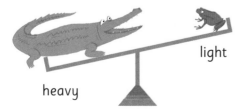

heavy / light

metre A unit for measuring the length of larger objects.

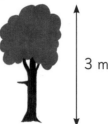

3 m

month The time it takes the moon to orbit the Earth. There are 12 months in a year.

January February March
April   May   June
July   August   September
October November December

near doubles A way to add two nearly identical numbers by using known doubles facts.

Example: 4 + 5 = 4 + 4 + 1 = 9

number line A line on which numbers can be placed to show their order in our number system or to help with calculations.

number sentence A way to record calculations using numbers and mathematical symbols.

Example: 23 + 7 = 30

numeral A figure or symbol used to represent a number.

Example:

1 – one   2 – two   3 – three

octagon A 2D shape with 8 sides.

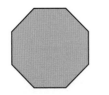

ordinal numbers Numbers that show the order or position of something in relation to others.

1st 2nd 3rd 4th 5th 6th

pair Two items that go together.

Example: Pairs that make 4

2 and 2   3 and 1

Pair of socks

parallel lines Straight lines that are the same distance apart and so will never cross.

parallel   parallel   not parallel

partitioning Dividing or separating an amount into parts.

Example: Some of the ways 10 can be partitioned are:

5 and 5   4 and 6   9 and 1

pattern A repeating design or sequence of numbers.

Example: Shape pattern

Number pattern

2, 4, 6, 8, 10, 12

pentagon A 2D shape with 5 sides.

pictograph A way of representing data using pictures to make it easy to understand.

Example: Favourite juices in our class

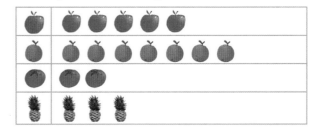

place value The value of a digit depending on its place in a number.

Hundreds	Tens	Ones
		8
	8	6
8	6	3

position Where something is in relation to other items.

Example: The boy is under the tree that is next to the house.

prism A 3D shape with parallel bases of the same shape and rectangular side faces.

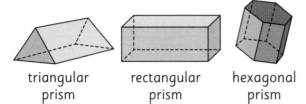

triangular prism rectangular prism hexagonal prism

pyramid A 3D shape with a 2D shape as a base and triangular faces meeting at a point.

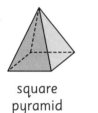 

square pyramid hexagonal pyramid

quadrilateral Any 2D shape with four sides.

quarter One part of a whole or group divided into four equal parts. Also used in time for 15 minutes.

Example:

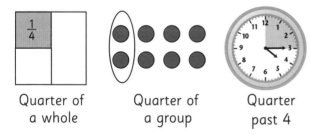

Quarter of a whole Quarter of a group Quarter past 4

rectangle A 2D shape with four sides and four right angles. The opposite sides are parallel and equal in length.

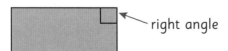 right angle

rhombus A 2D shape with four sides, all of the same length and opposite sides parallel.

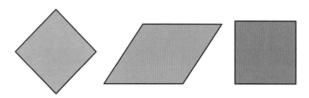

skip counting Counting forwards or backwards by the same number each time.

Example: Skip counting by 5s: 5, 10, 15, 20, 25, 30

Skip counting by 2s: 1, 3, 5, 7, 9, 11, 13

slide To move a shape to a new position without flipping or turning it. Also known as *translate*.

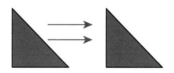

sphere A 3D shape that is perfectly round.

split strategy A way to solve number problems that involves splitting numbers up using place value to make them easier to work with.

Example: 21 + 14 = 35

square A 2D shape with four sides of equal length and four right angles. A square is a type of rectangle.

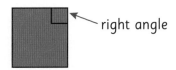

strategy A way to solve a problem. In mathematics, you can often use more than one strategy to get the right answer.

Example: 32 + 27 = 59

Jump strategy

Split strategy

30 + 2 + 20 + 7 = 30 + 20 + 2 + 7 = 59

subtraction The taking away of one number from another number. Also known as *subtracting*, *take away*, *difference between* and *minus*.

Example: 5 take away 2 is 3

survey A way of collecting data or information by asking questions.

Strongly agree	☐
Agree	✓
Disagree	☐
Strongly disagree	☐

table A way to organise information that uses columns and rows.

Flavour	Number of people
Chocolate	12
Vanilla	7
Strawberry	8

tally marks A way of keeping count that uses single lines with every fifth line crossed to make a group.

three-dimensional or 3D A shape that has three dimensions – length, width and depth. 3D shapes are not flat.

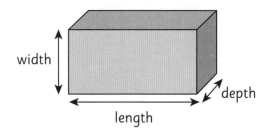

trapezium A 2D shape with four sides and only one set of parallel lines.

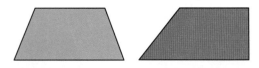

triangle A 2D shape with three sides.

turn Rotate around a point.

two-dimensional or 2D A flat shape that has two dimensions – length and width.

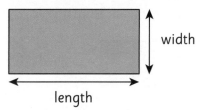

unequal Not having the same size or value.

Example:

Unequal size   Unequal numbers

value How much something is worth.

Example:

This coin is worth 5c.   This coin is worth $1.

vertical At a right angle to the horizon or straight up and down.

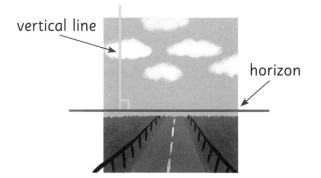

volume How much space an object takes up.

Example: This object has a volume of 4 cubes.

week A period of time that lasts 7 days.

Monday   Tuesday   Wednesday

Thursday   Friday   Saturday   Sunday

whole All of an item or group.

Example:

A whole shape   A whole group

width How wide an object is from one side to the other.

Example: This poster is 2 pens wide.

year The time it takes the Earth to orbit the Sun, which is approximately 365 days.

ANSWERS

Please note that where multiple answers to a question are possible, the most likely answers have been given as a guide.

UNIT 1: Topic 1

Guided practice
1 **a–b** Teacher to check. Teacher: Look for answers that show ability to start at the correct place to form numbers, and to follow the lines of each number accurately.

Independent practice
1 **a–b** Teacher to check. Teacher: Look for answers that show ability to accurately copy the numbers, and check for evidence of correct starting points as students write numbers independently.

2 **a** 0 1 2 **3 4** 5 **6 7 8** 9 **10**
  b 10 9 8 **7** 6 **5 4** 3 **2 1 0**
  c 4 5 6 **7 8 9** 10

3 **a** 2 **b** 3 **c** 5 **d** 6
  e 8 **f** 10

4 **a** 1 **b** 3 **c** 4 **d** 6
  e 8 **f** 9

5 **a** 1, 3 **b** 7, 9 **c** 4, 6 **d** 6, 8
  e 5, 7 **f** 8, 10

Extended practice
1 **a** Carriages numbered from left to right: 0, 1, 2, 3, 4.
  b Teacher to check. Teacher: Look for answers that show ability to read and interpret the numbers in order to draw the correct number of people in each carriage.

2 **a**

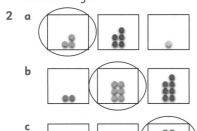

UNIT 1: Topic 2

Guided practice
1 **a**

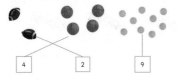

b

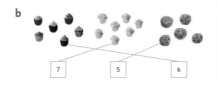

Independent practice
1 **a** 8 **b** 2 **c** 9 **d** 3
  e 5 **f** 10 **g** 4 **h** 6

2 **a–f:** Teacher to check. Teacher: Look for answers that show ability to read and interpret the numbers correctly and draw the corresponding number of items.

Extended practice
1 **a** 3 **b** 8 **c** 4

2 **a–c:** Teacher to check. Teacher: Look for answers that show ability to read and interpret the numbers correctly and draw the corresponding number of dots.

3 **a** 3 4 5 7
  b 1 6 8 10

UNIT 1: Topic 3

Guided practice
1 **a** 6 & 5 **b** No
2 **a** 7 & 7 **b** Yes

Independent practice
1 **a** 5 **b** 7 **c** 10 **d** 8 **e** 3

2 **a–h:** Teacher to check. Teacher: Look for answers that show ability to draw the correct number of items to match the given totals.

Extended practice
1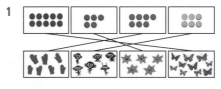

2 **a** 8 **b** 9
3 **a** 5 **b** 7

UNIT 1: Topic 4

Guided practice
1 **a** 3 **b** 1 **c** 4 **d** 5 **e** 2

Independent practice
1 **a**

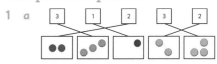

NOTE: There are two possible ways to match the 2 and the 3. Either is correct.

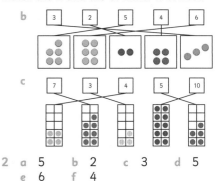

2 **a** 5 **b** 2 **c** 3 **d** 5
  e 6 **f** 4

3 **a** The third group should be circled.
  b The second group should be circled.

Extended practice
1

2

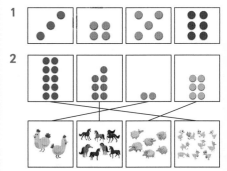

UNIT 1: Topic 5

Guided practice
1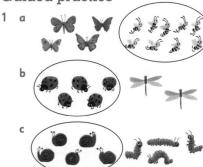

Independent practice
1 **a** 8 **b** Teacher to check. Teacher: Look for answers that demonstrate students' understanding of "more" by drawing more than 8 eggs.

2 a 6   b Teacher to check.
   Teacher: Look for answers that demonstrate students' understanding of "more" by drawing more than 6 chocolates.
3 a 7   b Teacher to check.
   Teacher: Look for answers that demonstrate students' understanding of "less" by drawing less than 7 socks.
4 a 4   b Teacher to check.
   Teacher: Look for answers that demonstrate students' understanding of "less" by drawing less than 4 toys.
5 a Same   b Less   c Same
   d More

Extended practice

1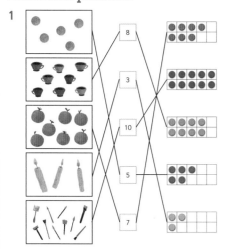

UNIT 1: Topic 6

Guided practice

1
2 a red   b grey   c green

Independent practice

1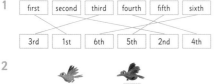
2
3 first, second, third, fourth
4 a cat   b cow   c dog   d frog
5 a

b
c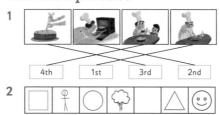

Extended practice

1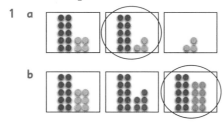
2

UNIT 1: Topic 7

Guided practice

1 a–c Teacher to check. Teacher: Look for answers that show students' ability to start at the correct place to form their numbers, and to follow the lines of each number accurately.

Independent practice

1 a–b Teacher to check. Teacher: Look for answers that show ability to accurately copy the numbers from the box above, and check for evidence of correct starting points as students write numbers independently.
2 a 12, 15, 16, 19, 20
   b 19, 15, 14, 11   c 15, 18, 19
3 a 11   b 14   c 16   d 17
   e 19   f 20
4 a 10   b 12   c 9   d 15
   e 18   f 19
5 a 13, 15   b 16, 18   c 14, 16
   d 11, 13   e 18, 20   f 10, 12

Extended practice

1 a
   b
2 a 14, 15, 17, 19, 22, 26, 28
3 a 10, 11   b 19, 20   c 22, 23
   d 28, 29   e 17, 18   f 27, 28

UNIT 1: Topic 8

Guided practice

1 a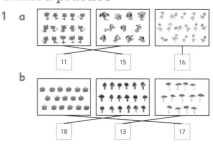
   b

Independent practice

1 a
   b
2 a 12   b 15   c 14   d 17
3 a 10   b 5   c 15
4 a 10   b 3   c 13
5 a 10   b 9   c 19
6 a 10   b 6   c 16

Extended practice

1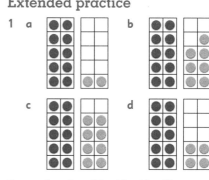
2 a 2   5   12   14   16   19
   b 6   7   10   11   17   20
   c 1   2   5   11   12   15

UNIT 1: Topic 9

Guided practice

1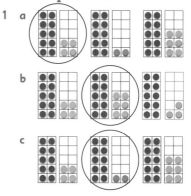

OXFORD UNIVERSITY PRESS   125

d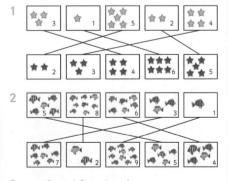

Independent practice

1   a–b   Teacher to check. Teacher: Look for answers that demonstrate understanding of the concept of "more". Students should also be able to count accurately to draw more than 7 and 14 items respectively.
2   a–b   Teacher to check. Teacher: Look for answers that demonstrate understanding of the concept of "less". Students should also be able to count accurately to draw less than 15 and 17 items respectively.
3   a   The planes should be circled.
    b   2
4   a   The rainbows should be circled.
    b   3
5   a   The socks should be circled.
    b   2

Extended practice

1   a   Yes   b   Yes   c   No   d   No
2   a
    b   14
3   a
    b   29

UNIT 1: Topic 10

Guided practice

1   a   0 and 5 makes 5
    b   1 and 4 makes 5
    c   2 and 3 makes 5
    d   3 and 2 makes 5
    e   4 and 1 makes 5
    f   5 and 0 makes 5

Independent practice

1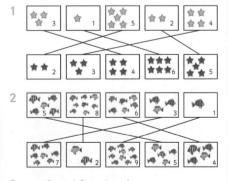

2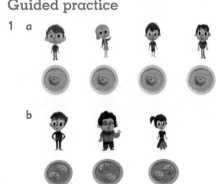

3   a   3 and 3 makes 6
    b   3 and 4 makes 7
    c   5 and 5 makes 10
    d   4 and 6 makes 10
    e   2 and 8 makes 10

f   7 and 3 makes 10
g   9 and 1 makes 10
h   1 and 9 makes 10

Extended practice

1   a   2   b   5   c   4   d   9
2   a–b   Teacher to check. Teacher: Look for answers that show students' ability to accurately partition 10 into two parts. Students' number sentence should also match the visual representation of the sum.

UNIT 1: Topic 11

Guided practice

1   a   4   b   2
2   a   6   b   2

Independent practice

1   a
    b
    c
    d
    e

2   a   Unequal   b   Equal
    c   Equal   d   Unequal

Extended practice

1–3   a–b   Teacher to check. Teacher: Look for answers that show ability to make equal groups independently, and to correctly identify the number of items in each group.

UNIT 1: Topic 12

Guided practice

1   a
    b

Independent practice

1   a
    b   4
2   a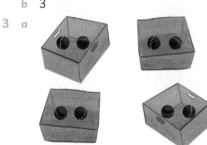
    b   3
3   a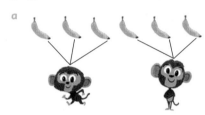
    b   2
4   a
    b   3 (NOTE: Students may draw the lines in a different order – as long as they share 3 bananas to each monkey this is fine.)
5   a
    b   2 (NOTE: Students may draw the lines in a different order – as long as they share 2 muffins to each child this is fine.)
6   a
    b   3

Extended practice

1–2   a–b   Teacher to check. Teacher: Look for answers that show ability to make equal groups independently and to correctly identify the number of items in each group.
3   a and d should be ticked.

UNIT 2: Topic 1

Guided practice

1 a–d Only one half of each shape should be coloured in.

Independent practice

1 a Whole   b Half   c Half
   d Whole   e Whole   f Half

2 a

   b

   c

   d

Extended practice

1 a

   b

   c

   d

   e

   f

2 Students could divide the square in any 2 of the following ways:

UNIT 3: Topic 1

Guided practice

1 The following places should be ticked: an the movies; at the supermarket; at an ice-cream shop; in a restaurant. Allow variations in answers if students can offer justifications:

e.g. "I need money walking in the park because there is a kiosk where I buy an ice-cream."

Independent practice

1 15

2 a 10   b 4

3 a 14   b 12   c Flynn

4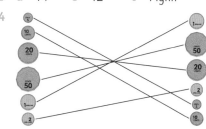

Extended practice

1 Students' own answers. Look for students who can choose appropriate scenarios where money is likely to change hands and who understands the transactional nature of money.

2 a 5 coins should be circled.
   b 2 coins should be circled.
   c 8 coins should be circled.

UNIT 4: Topic 1

Guided practice

1 a–c

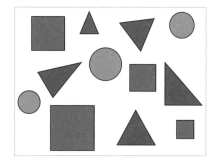

2 a 4   b 5   c 3

Independent practice

1 a–b

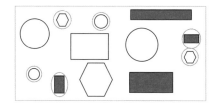

2 a–c

3 a
   b
   c

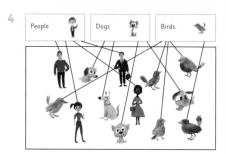

4

Extended practice

1 Teacher to check. Teacher: Look for answers that show ability to choose appropriate groupings, such as identifying the colours of the particular foods or differentiating the fruits from the vegetables.

2 Teacher to check. Teacher: Look for answers that show ability to choose appropriate categories such as size, shape or colour, and to successfully sort the shapes based on the categories identified.

UNIT 4: Topic 2

Guided practice

1 a
   b
   c
   d

Independent practice

1 a
   b
   c
   d
   e

2 a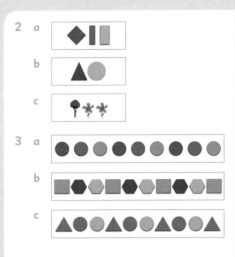

  b

  c

3 a

  b

  c

Extended practice

1 **a–b** Teacher to check. Teacher: Look for answers that show ability to make an identifiable pattern using the given shapes, with or without the aid of colour patterning.

2 a

  b

UNIT 4: Topic 3

Guided practice

1 Teacher to check. Teacher: Look for answers that demonstrate an understanding of patterns by correctly drawing a pattern with repeating elements in sequence.

Independent practice

1–4 It does not matter where in the sequence the repeating section is circled, as long as the student correctly identifies the repeating elements.

1 a A sun and 2 moons should be circled. b 3

2 a A heart, a star and a flower should be circled. b 3

3 a A cloud and a lightning bolt should be circled. b 2

4 a A rainbow, sun, cloud and moon should be circled. b 4

5 a

  b

6 a

  b

7 a The second option with the hexagon and the cube should be circled.

  b The first option with the inverted triangle and the rectangle should be circled.

Extended practice

1 **a–b** Teacher to check. Teacher: Look for answers that show ability to follow the directions to make a pattern with the given number of repeating elements.

2 **a–c**

  d 3

UNIT 5: Topic 1

Guided practice

1 a

  b

2 a

  b

Independent practice

1 a Teacher to check. Teacher: Look for answers that show students' understanding of "longer" by drawing a line that is longer than the one on the page.

  b Teacher to check. Teacher: Look for answers that show students' understanding of "shorter" by drawing a line that is shorter than the one on the page.

2 a Teacher to check. Teacher: Look for answers that show students' understanding of "taller" by drawing a building that is taller than the one on the page.

  b Teacher to check. Teacher: Look for answers that show students' understanding of "shorter" by drawing a building that is shorter than the one on the page.

Guided practice

1 a   b

  c   d

Independent practice

1 **a–c** Teacher to check. Teacher: Look for answers that show ability to cover the front of the chosen books without gaps, accurately count the blocks used and use this information to identify the one with the greater area.

2–3 Teacher to check. Teacher: Look for answers that show ability to select objects that have an area of approximately 8 and 12 blocks, and to then measure the area using the blocks with no gaps between them.

Extended practice

1 **a–c** Teacher to check. Teacher: Look for answers that show students' ability to make credible choices, and to justify why they chose particular items.

2 **a–b** Teacher to check. Teacher: Look for answers that show students' ability to make reasonable guesses as to items with larger or smaller areas than the book, and to justify their responses.

UNIT 5: Topic 2

Guided practice

1 a b

2 a   b

Independent practice

1 a Teacher to check. Teacher: Look for answers that show students' understanding of "takes up more space" by drawing an item that has a greater volume than the block of chocolate.

b Teacher to check. Teacher: Look for answers that show students' understanding of "takes up more space" by drawing an item that has a greater volume than the loaf of bread.

2 a–b Teacher to check. Teacher: Look for models that show students' ability to use the correct number of blocks.

c The "T" shape should be circled.

Guided practice

1 a   b

2 a   b

Independent practice

1 a Teacher to check. Teacher: Look for answers that show students' understanding of "holds more" by drawing an item that has a greater capacity than the mug, such as a juice bottle or bucket.

b Teacher to check. Teacher: Look for answers that show students' understanding of "holds more" by drawing an item that has a greater capacity than the bath, such as a lake or swimming pool.

2

Extended practice

1 a–c Teacher to check. Teacher: Look for answers that show students' ability to use strategies such as pouring from one container to another to determine which has the greater capacity, and to then correctly classify their two containers according to capacity.

2 a–c Teacher to check. Teacher: Look for answers that show students' ability to fill each box with blocks without leaving gaps, and to accurately count and compare the volume of the two boxes to correctly classify them according to volume.

UNIT 5: Topic 3

Guided practice

1 a   b

2 a   b

Independent practice

1 a   b

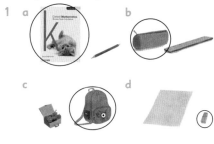

c   d

2 a

b Answers will vary, depending on whether the lunch box and water bottle used are empty or full.

c   d

3 a Heavy b Light c Heavy

4 a The apple should be coloured in.
   b The watch should be coloured in.

5 a The car should be coloured in.
   b The couch should be coloured in.

Extended practice

1 a–b Teacher to check. Teacher: Look for answers that show ability to identify items heavier than the given items and justify choices using appropriate language such as "lighter" and "heavier".

2 a–b Teacher to check. Teacher: Look for answers that show ability to identify items lighter than the given items and justify choices using appropriate language such as "lighter" and "heavier".

3 a–b Teacher to check. Teacher: Look for answers that show students' ability to make plausible guesses about objects that are easy or hard to push or pull, and that justify answers using appropriate language.

UNIT 5: Topic 4

Guided practice

1 a

b

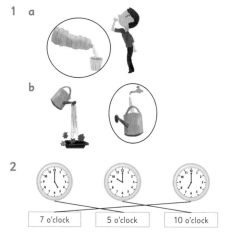

2

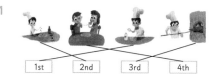

Independent practice

1

2 NOTE: Teacher to check. The answers below are a guide only. Teacher: Accept any reasonable answers so long as students can justify their choices (e.g. "We play a board game at home that is really quick.").

a

b

3 a 4 o'clock   b 1 o'clock
   c 6 o'clock   d 9 o'clock
   e 11 o'clock f 7 o'clock

4

Extended practice

1 a–b Teacher to check. Teacher: Look for answers that show students' ability to choose appropriate activities based on their understanding of morning and afternoon, and to give reasonable justification for the timing of these activities.

2 a–c Teacher to check. Teacher: Look for answers that show students' ability to correctly show "o'clock" time, and to make an accurate estimate to the nearest hour of when they do everyday activities.

UNIT 5: Topic 5

Guided practice

1. a
| Monday |
| Tuesday |
| Wednesday |
| Thursday |
| Friday |
| Saturday |
| Sunday |

b Teacher to check. Teacher: Look for answers that show students' ability to write their favourite day correctly, and to offer logical reasons for why it is their favourite.

Independent practice

1.
| Monday | Tuesday | Wednesday | Thursday | Friday | Saturday | Sunday |

2. a A line should be joined to Monday, Tuesday, Wednesday, Thursday and Friday.
   b Answers will vary.
   c Answers will vary.
   d A line should be joined to Sunday.

3. a–d Teacher to check. Teacher: Look for answers that show ability to accurately identify activities that occur on particular days.

4. a–c Teacher to check. Teacher: Look for answers that show students are aware of the days of the week and can identify the day before and after the current day.

Extended practice

1. a
| Monday | Tuesday | Wednesday | Thursday |

   b
| Saturday | Sunday | Monday | Tuesday |

2. a Wednesday   b Saturday
   c Friday   d Monday

3. a Sunday   b Friday
   c Tuesday   d Thursday

UNIT 6: Topic 1

Guided practice

1. a–b

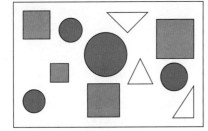

2. a–b

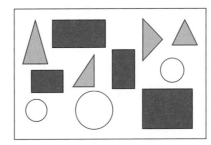

Independent practice

1.

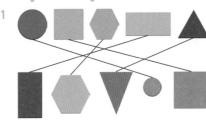

2. a–b

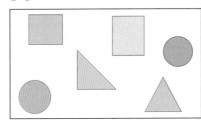

3. a 4 corners 4 sides
   b 0 corners 0 sides
   c 3 corners 3 sides
   d 4 corners 4 sides

4. a 4   b 5   c 3

Extended practice

1. a–c Teacher to check. Teacher: Look for answers that show an understanding of corners and sides, and an ability to draw shapes that meet the given criteria.

2. a 8   b 11   c 3

UNIT 6: Topic 2

Guided practice

1. a–b

2. a–b

Independent practice

1.

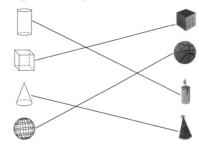

2. a Students should have drawn a square.
   b 6

3.

4. a 4   b 3   c 2   d 2

Extended practice

1. a

   b

   c

2. a Teacher to check. Teacher: Look for answers that show ability to identify and represent common items that are cubes, such as dice.
   b Teacher to check. Teacher: Look for answers that show ability to identify and represent common items that are spherical, such as sports balls.

UNIT 7: Topic 1

Guided practice

1. a under   b towards
   c above   d between

Independent practice

1. a–f Teacher to check. Teacher: Look for answers that show ability to correctly interpret positional language to place the ball as instructed.

2. a left   b right   c left

3

Extended practice
1 Teacher to check. Teacher: Allow variations on given answers if students accurately describe the location of the objects.
  a above   b on   c behind
2 a True   b False   c False

UNIT 7: Topic 2

Guided practice
1 a–c Teacher to check. Teacher: Look for answers that show students' ability to draw a direct route between the two locations, staying on the path, and to explain how they chose their route.

Independent practice
1 a your classroom   b Year 4
  c Year 3
2 a the parking area   b the garden
  c the hall
3 a the library   b the office
  c Year 5
4 a False   b True   c True
  d False   e True
5 a–c the playground

Extended practice
1 a–d Teacher to check. Teacher: Look for answers that show students' ability to draw a direct path using two different routes from the door to their favourite location, and to accurately identify the other locations they pass on the way.

UNIT 8: Topic 1

Guided practice
1 a–e Teacher to check. Teacher: Look for answers that show students' ability to identify the option that applies to them.

Independent practice
1–3 a–c Teacher to check. Teacher: Look for answers that show students' ability to find answers from exactly 5 students and record them accurately. Students should also be able to identify how many of each response they received.
4 a 5   b 3   c yes
5

| Yes | ☺ | ☺ | ☺ | ☺ | ☺ |
| No | ☹ | ☹ | ☹ | ☹ | ☹ |

a 3   b 2

Extended practice
1 a–b Teacher to check. Teacher: Look for answers that show ability to use the question scaffolds to write 2 questions that could be posed to gain data.
2 a Teacher to check. Teacher: Look for answers that show students' ability to choose a question that is appropriate for their classmates.
  b Teacher to check. Teacher: Look for answers that show students' ability to pose their question to 5 students and record the responses accurately.
  c Teacher to check. Teacher: Look for answers that show students' ability to use their collected data to colour the faces accurately.

UNIT 8: Topic 2

Guided practice
1 a 5   b 7   c 2   d 3

Independent practice
1 a
  b
  c Yes
2 a

| Yes | ☺ | ☺ | ☺ | ☺ | ☺ | ☺ |
| No | ☹ | ☹ | ☹ | ☹ | ☹ | ☹ |

b Yes   c 3   d 5   e 8

Extended practice
1 a–b Teacher to check. Teacher: Look for answers that show students' ability to recognise how to record the information given to them by their classmates. Students should also have 10 cells of the graph coloured, to represent 10 students.
2 a–e Teacher to check. Teacher: Look for answers that show students' ability to accurately interpret their pictograph.